THE 12 SIMPLE SECRETS OF MICROSOFT MANAGEMENT

THE 12 SIMPLE SECRETS

OF MICROSOFT

MANAGEMENT

HOW TO THINK AND ACT
LIKE A MICROSOFT MANAGER AND
TAKE YOUR COMPANY TO THE TOP

DAVID THIELEN

with Shirley Thielen

MCGRAW-HILL

NEW YORK   SAN FRANCISCO   WASHINGTON, D.C.   AUCKLAND   BOGOTÁ
CARACAS   LISBON   LONDON   MADRID   MEXICO CITY   MILAN
MONTREAL   NEW DELHI   SAN JUAN   SINGAPORE
SYDNEY   TOKYO   TORONTO

This book is dedicated
to our three precious daughters,
Winter Maile, Tanya Nicole, and Brianna Leilani.
They are a gift beyond measure
and more precious than life.
We love you.

McGraw-Hill

A Division of The *McGraw-Hill* Companies

2 3 4 5 6 7 8 9 0   DOC/DOC   9 0 4 3 2 1 0 9

ISBN 0-07-134248-6

This publication is designed to provide accurate and authoritative information in regard to the subject matter covered. It is sold with the understanding that the publisher is not engaged in rendering legal, accounting, or other professional service. If legal advice or other expert assistance is required, the services of a competent professional person should be sought.

—From a declaration of principles jointly adopted by a committee of the American Bar Association and a committee of publishers.

Printed and bound by R. R. Donnelley & Sons Company.

Trademarks
DOS, MS-DOS, Windows, Windows 3.0, Windows NT, Internet Explorer, Office, Word, Excel, Microsoft, and MS Money are trademarks of Microsoft Corporation. IBM and OS/2 are trademarks of International Business Machines. Quicken is a registered trademark of Intuit, Inc. Jeep Cherokee is a registered trademark of Chrysler Corporation. All trademarks and copyrights mentioned are the properties of their respective owners.

McGraw-Hill books are available at special quantity discounts to use as premiums and sales promotions, or for use in corporate training programs. For more information, please write to the Director of Special Sales, McGraw-Hill, 11 West 19th Street, New York, NY 10011. Or contact your local bookstore.

 This book is printed on recycled, acid-free paper containing a minimum of 50% recycled, de-inked fiber

CONTENTS

THE EMPEROR'S NEW CLOTHES

(with apologies to Hans Christian Andersen)

We're all familiar with Hans Christian Andersen's story "The Emperor's New Clothes." In our version, the emperor is the CEO of a major corporation. Consultants (the weavers) convince the CEO that they know how to weave the most beautiful cloth imaginable. Not only are the colors and pattern unusually beautiful, but clothes sewn from this special hi-tech material have the remarkable quality of being invisible to anyone who is not good at his job or who is incredibly stupid.

At the annual company meeting the CEO, accompanied by his entourage of vice presidents (strains of "Chariots of Fire" playing in the background), steps into the spotlight wearing his new suit—i.e., absolutely nothing. An entry-level employee says loudly, "But he hasn't got anything on!" Stunned silence ensues, followed by a rush of activity.

Building security surrounds the employee and escorts him at walkie-talkie point out of the meeting. After fervent whispering into the radios, the directive is received to bring the employee to the human resources office. Duly noting that the employee's badge is $\frac{1}{4}$ inch out of the prescribed alignment, a serious violation of company policy, security leaves the office, confident in a job well done. Meanwhile, back at the meeting, a recess has been called. During the recess the CEO and his vice presidents express concern to one another about the psychotic episode just witnessed and discuss what steps could be taken to ensure that further episodes do not occur. The meeting is called to order, with the announcement that mental health coverage is not included under the company's insurance plan.

THAT AFTERNOON

The employee's manager files a written warning, admonishing the employee and officially notifying him that disruptions of this nature will not be tolerated and future episodes will result in immediate termination. The warning is delivered to the employee with three members of the human resources staff in attendance as witnesses.

THE NEXT DAY

Numerous closed-door meetings are held throughout the company. An off-site management retreat is scheduled. The topic is rumored to be "Regaining Control."

The employee is enrolled in an attitude-adjustment seminar. His manager is registered for a class on "managing difficult people," because clearly the manager isn't capable of controlling his direct reports, with a note made in his personnel file outlining this deficiency.

TWO DAYS LATER

A three-page memo from the executive office outlining proper meeting procedures is circulated to all employees. Middle managers are advised to attend a conference on "dressing for success."

THE NEXT WEEK

Team meetings are held throughout the company detailing the fabulous properties of the special cloth. Each team is tasked with determining how to leverage the cloth's qualities to increase market share.

TWO WEEKS LATER

The consultants are granted a long-term contract, complete with company car and expense accounts.

The senior executives are all given huge bonuses for their intelligence and suitability to their jobs.

A MONTH LATER

The CEO has the employee fired. The termination paperwork states the reasons for the dismissal as insubordination and incompetence.

FOREVER AFTER

The CEO continues to wear suits made by the consultants at meetings, where everyone compliments him on the fine fabric. The employee tried to find a new job in the same industry, but no company was willing to hire such a disruptive troublemaker. He now makes telemarketing calls—during your dinner hour.

Too bad he turned down that internship with Microsoft . . .

FEEL THE ENERGY

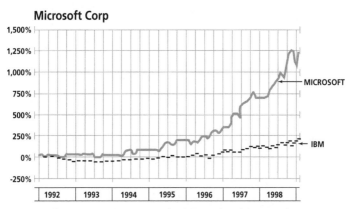

The return on investment for Microsoft and IBM stock.

O ver the last 20 years Microsoft has grown from two friends in a dorm room to the second most highly capitalized company in the United States. This has occurred not because of luck, money, acquisitions, or personal connections. It has occurred solely because Bill Gates built a company with his own unique management style.

Microsoft's management style is its core strength. There are other companies that produce better software, market better, and make fewer mistakes, but no other large company manages its business as well. Microsoft's management strength is why it will continue to dominate.

And the juggernaut shows no signs of slowing down. Microsoft continues to increase its gross and net profits by 40 percent or more each year. It increases its income by devouring new market after new market. Microsoft has pulled off the impossible by retaining the growth levels of a small company as it becomes one of the largest companies in the world.

THE COMPETITIVE THREAT

Microsoft's strength directly or indirectly threatens every company on the globe. If Microsoft sets its sights on another company's business, that company can, at best, hold on to part of its former market. At worst, it will have to close its doors.

In markets that Microsoft itself does not pursue, companies that have learned valuable lessons from Microsoft will go after those markets. Other companies are learning from Microsoft and are using these lessons to dominate their own markets.

Microsoft has caused a change in the business community greater than the change due to global competition that has occurred over the last 20 years. When markets like cars and electronics became global, companies had to make significant changes to survive. But the changes were possible—cut some of the fat and pay more attention to quality. (It helped also that governments could step in and slow down the transition through the use of trade barriers.)

Yet most companies do not see it coming. They see that Microsoft is successful, but they don't understand that Microsoft's methods directly threaten them. When Microsoft, or a company managed like Microsoft, does decide to go after another company's business, it's too late to change.

This bears repeating. If a company is still managing itself the old way when Microsoft goes after that company's markets, it's too late for the company to change. It will be in severe financial trouble in one to three years and out of business in two to five years. This is the threat faced by businesses that refuse to change their management practices.

The change required is wrenching for most corporations. It means a change in the mind-set of every manager and employee within the corporation. This change is required because Microsoft has every one of its employees 100 percent focused on dominating each market it targets.

Companies need to evolve or die. They need to

evolve now, today, before Microsoft or another competitively managed company goes after their markets. The days of not solving problems until they are obvious are over. By the time the problem becomes obvious, there is not enough time left to fix it.

The fact that some of the concepts covered herein are unusual and the benefits may not be immediately obvious is not a reason to ignore them. To even survive the competition of a Microsoft, much less thrive against it, you must understand how critical these concepts are. Without these concepts you are using a bow and arrow against a tank. You can fight bravely and wisely. You may even win some battles. But you will always lose the war.

> *The world has changed. You can change with it, or you can be collateral damage when your company is destroyed. Like it or not, you're in a war, and opting out is not an option. If you choose not to fight, the only question then is when will a new-line company set its sights on your market . . . and destroy you.*

MAKING THE CHANGE

OK, so you're convinced. You're ready to turn your company inside out to make it competitive in this new environment. You intend not just to survive but to excel in this new business environment. You're ready to be one of the few companies that is taking market share away from the many failing companies.

But . . . what should you do? Well, oddly enough, this book tries to answer that question. This book describes Microsoft's management style and corporate culture as it applies to managing a company. It does not discuss how to manage a software company—it discusses how to manage any company the same way as Microsoft does.

Equally important, it does not describe only those parts of Microsoft's system that the author agrees with. It paints a complete portrait of Microsoft's management and culture as it is practiced. It explains the secrets of Microsoft's success and what other companies must do to emulate that success.

FEEL THE ENERGY

Walk into a building at Microsoft. Any building. It's in the air. You can feel the same energy you feel when you're playing on a winning sports team with the spectators cheering you on. This is not pseudoscience mumbo-jumbo. Most visitors at Microsoft feel it and comment on it one way or another.

We leave Microsoft and walk into a building down the street. It's another company. And we feel . . . nothing. How do you describe that other company? It's dead. By this measure Microsoft is very alive, as are a few other companies. Some companies are alive, but nowhere near as vibrant. Sadly, most companies are dead.

A sports team going through the motions will never beat an inspired team. A dead company will never beat one that is alive. At its most elementary level, this is the fundamental issue facing all of us. To succeed, a company must be alive. Yet most managers don't even understand this issue, much less address it.

THE 12 SECRETS

This book will guide you through the 12 basic secrets of Microsoft's method—secrets that are applicable across a wide range of industries. Will implementing these 12 basic elements guarantee you success? No, because your competitor may have implemented them better than your company did. Should you implement these 12 basic elements exactly as laid out in this book? No, because each company and situation is different. You need to use this book as a guide, not as a straitjacket.

All we can promise is that if you don't understand and implement these 12 basic elements, and your competitors do, your chances of failure are extremely high.

The good news is that each of these elements is easy to implement in practice. The bad news is that most of them are, at many companies, politically impossible to implement. So keep in mind that you personally have two basic options. One is to make your company ready for competition at this level. The other is to switch companies and go to work for a company that has already made itself ready.

TOTAL
WORLD DOMINATION

Microsoft is going after 100 percent of every market it is in.

—Mike Maples, when he was executive vice president

A group of us were talking at lunch one day. Some of the Microsoft cafeterias have tables outside, and we were enjoying one of the few sunny days Seattle has to offer. One of the people at the table commented on the fact that Microsoft has no corporate motto (using the term *mission statement* at Microsoft is a good way to lose all credibility). Another person at the table offered up the opinion that there was a corporate motto: it just wasn't official. Jokingly, three of us said the same thing at the same time—"Total World Domination."

Here's the critical point: I know for certain I had never heard this statement before, either seriously or as a joke. I don't think the others had either. Yet three of us said it at the same time and everyone else at the table agreed.

All Microsoft employees know in their gut what their primary goal is. And that is to win 100 percent of whatever market they are going after. Every employee at Microsoft, in every decision made, in every meeting, in every product design, has the same basic goal.

And when various groups at Microsoft need to work together, this common goal encourages them to help each other rather than to work at cross-purposes. Because only by helping other groups as needed can the company achieve total domination.

All of the employees at your company have to know, in their gut, that the primary goal is *total world domination*. Whatever their job, their primary task is to perform that job in such a way that your company can own 100 percent of your market—99 percent is not acceptable. And they all need to fight as hard to take that last 1 percent as they fought to take the first 10 percent. Implementing this is simple. In every review, the first and most important question should be, "What are you doing to increase market share?" And then drill down. What else should be done? What can be done better?

Make sure that you are getting well-thought-out answers. Responses will be weak at first. But over time they will improve, as long as you stay on your employees and demand better solutions each time. Pretty soon, because you are always demanding better answers, your employees will constantly be in the mind-set of determining how to increase market share.

CONQUEST

> *"We didn't write Excel to make money, we wrote it for the sheer joy of putting the largest computer software company out of business."*
> *(Lotus was bigger than Microsoft at the time.)*
>
> — Chris Peters, when he was head of Excel development

Psychology says that human beings have several levels of needs and one level has to be satisfied before people worry about the next level. At the base are food and shelter. For most people, a couple of levels above that is the need for security.

However, for some, the primary need at that level is conquest, not security. These people have a very different view of life. In business, people who want security are satisfied with a sufficient profit that provides security. Those driven by conquest, however, are only satisfied with 100 percent market share, even if their net profit is smaller.

Most businesses are managed by people who value security well over conquest. Security requires a reasonable profit, so making a reasonable profit is important to most managers.

Increasing profits increases security. So increasing profits again is important to most managers. But not that important. Increasing profits is not worth risking the reasonable profits already being made because that risks security—and security is not to be risked.

Microsoft takes the opposite tack. Risking profits to increase market share is well worth it. After all, losing profits merely threatens security, while gaining market share shows who won. Another way to put it is that Microsoft is composed primarily of type A overachievers desperate to prove that they are the best. And they want to prove this not just by doing better than the competition, but by destroying it.

At the same time, do not confuse random, rash acts

(*bad*) with bold, measured risks (*good*). Recklessness will destroy a company just as quickly as caution. Encourage the bold, not the reckless.

The top priority in any company should be to hire type A people who are driven by conquest rather than security. It is critical that an organization have a core set of people who naturally gravitate toward conquest. But even Microsoft is not composed exclusively of natural-born risk takers, so for the remaining group you need to encourage this behavior. This can be accomplished in two ways:

> First, promote the people who are focused on increasing market share and who are willing to take risks to do so. This means you will promote those who make mistakes and not promote the overly careful employees who don't make mistakes. This both moves up the people who want to conquer their markets and forces security-conscious people who want to be promoted to take risks to advance their career.
>
> Second, fire the employees who are completely unwilling to take risks to increase market share. This shows security-conscious employees that their job is at stake if they always make the safe decision. This will force them to take risks in order to minimize risk.

EVERY STRATEGIC MARKET

Occasionally Microsoft will back out of a market. This can be due to a shrinking market, the competition not being worth the effort, or a myriad of other reasons. But it will never back out of a strategically critical market. Networking was added to Windows at no cost to the consumer to take the networking market away from Novell. Internet Explorer is provided for free to take the browser market away from Netscape. When Microsoft and IBM first broke up over OS/2, Steve Ballmer stated that Microsoft made more money when a copy of OS/2 was sold than when a copy of Windows was sold. He then went on to say that it was critical to have Windows beat OS/2 for Microsoft to retain its strategic position.

Microsoft is very comfortable losing money year after year to keep going after strategic markets. And while it doesn't win them all (for example, *Quicken* vs. *MS Money*), it does not ever give up.

What must you do? Determine your strategic markets. For each one work out how you are going to dominate it. Your plan will change over time, and that's fine. But stay focused on how to go about owning that market.

You may not be able to afford to go after every strategic market at once. So pick the most critical market and go after that one. You can learn a lot as you do. But make sure a critical priority at your company is dominating that strategic market.

Profits are secondary. Breaking even is good. Tying a strategic, but unprofitable, product to a moneymaking product so the combined product is lucrative is a tactic often used (and criticized by others) by Microsoft. But regardless of profits, the critical question must be: Are we successfully gaining market share?

If profits are secondary, does this mean a company will lose money? No. What it does mean is that by implementing the above guidelines, long-term profits will follow. Owning a strategic market gives you a cash cow.

LEAD BY EXAMPLE

The most critical element is the example set by management. Everything outlined above can be verbally stated (and repeated) to employees, but the unwritten rules, what is actually rewarded and punished, are the rules that are followed.

Keep the following rules in mind as you operate day to day: When an employee makes a mistake, if it was a reasonable mistake in pursuit of increasing market share, praise it and use it as an example to others of what you want them to do. In addition, ask your direct reports, daily at first and always at least weekly, what they are doing to increase market share.

And finally, when an employee refuses to take any risks to increase market share, demote that employee. A single demotion sends a very loud message to all employees that the safe route is no longer acceptable.

2

THE TOP 5 PERCENT

Several years ago I was scheduled to give a talk on managing software developers at a conference. Several days of the conference were devoted exclusively to the topic of managing the software development process. On the flight down, I was pulling together my list of what I had seen that was important at Microsoft when it hit me: The single most important contributor to productivity is the quality of the employees. Everything else is secondary to this one criterion.

This is the second of the two key management principles at Microsoft. Hire the very best people. No management system can make up for having less than the best employees, while superb employees can make up for lots of management stupidity (and Microsoft does make its share of dumb management decisions).

Yet no one even raises this subject. I have not read a single book that discusses the criticality of hiring excellent employees. And when I raise this issue in the talks I give at conferences, it is met with open hostility by many in the audience.

Why is this? Two explanations come to mind: First, it completely redefines the role of the manager and that is scary to most people in management. Second, the manager is no longer in control; the employees are.

AT THE MOVIES

Almost every year Disney releases another movie about a kid's sports team that is composed of losers. Their competition always has better players, better coaches, nicer uniforms, and better grooming, and the kids on the other teams always look cooler.

But the coach, after surmounting his [*his* is used because the coach is always a (white) male] own personal challenges, guides the kids on the team through the difficulties they face, and by overcoming those obstacles, they become a better team and win the championship, usually in a final one-on-one between the main character from each team. What is the basic assumption behind the story? That a coach can take a team and, through coaching (management), turn it magically from the worst to the best team in the league.

We all know this is pure fantasy, and yet we expect the same when it comes to management. A more instructive example from Hollywood is *The Bad News Bears*. In that particular movie Walter Matthau did what any intelligent coach would do when trying to improve a team: He recruited ringers. In short, he hired better employees.

A MESSAGE FROM REAL LIFE

The Denver Nuggets are beyond inadequate. Everyone knows it—and everyone knows why. It's because the players are woefully lacking in talent. (Compared with most of us they are excellent players. I'm speaking in relation to the other players in the NBA.) No one expects the team to win. In fact, everyone knows it can't win.

Even more to the point, the only shock in the NBA finals is if the Chicago Bulls lose a single game. They don't win because of superior coaching or nicer-looking uniforms. They win because of the playing skills of Michael Jordan, Scottie Pippen, etc. At the same time, the only way the Nuggets could win is to hire better players. We know that a team is only as good as its players.

If a coach said he didn't need the best players and his coaching could make up for a serious deficiency in player skills, that coach would shortly be looking for a new job.

With that said, the coach (manager) can, and should, contribute one very important thing: to build a sense of team spirit or esprit de corps to get the entire team working closely together. For business, just as in sports, the whole becomes so much more than the sum of the parts.

FALSE LOGIC

Walk into most any company and you will find the same system. The company hires people with the minimum qualifications for the job. And because pay scales are set as low as possible, the company generally ends up hiring people who barely meet those qualifications.

But that's OK. Because with good management these minimally qualified people can be directed to do an outstanding job. After all, its not the people doing the actual work that matter, just those who manage them.

And that's why there are so many books and seminars on management. Because obviously there must be a management secret to make any group better than its competition. It's just finding that silver bullet.

What a crock. The fundamental theory of business management, that managers determine the quality of the employees, is totally backward. Over the last 15 years corporate America has decimated the ranks of middle managers, frequently resulting in improved productivity. (Just imagine the improvement in productivity if upper management were equally reduced . . .)

Dictating to qualified teams of motivated and focused employees can only decrease the productivity of the team. The best managers are the ones that do the least damage to the teams they have working for them.

THE LOWEST COMMON DENOMINATOR

A large, established (i.e., old and tired) company acquired a successful start-up (names withheld to protect the guilty). There were teams in both companies working on comparable products. Therefore, the company did the "obvious" thing; it merged them.

Now the tired old company was smart. It realized that it was overly bureaucratic and its employees were not at all productive, so it physically placed the merged team at the start-up's location, which it had kept separate, so that only some of the bureaucracy transferred over.

The result? The quick-moving, effective teams from the start-up became hobbled by the people from the tired old company. Because half the team was now considerably less qualified than before, the best people started leaving out of frustration, and the remainder spent half their time trying to work around the average performers they had become saddled with.

Needless to say, over time the successful company was brought down to the effectiveness level of the tired old company (which then turned around to buy another company because it needed a success).

SEPARATING FACT FROM FICTION

I was discussing programming skills with a senior executive at a regional phone company. He was insisting that their programmers were among the best anywhere. The programmers there are required to follow a dress code, work in cubicles, and show up by exactly 9 a.m., all for a lot less money than they could get at a software company. Going in, the odds of having excellent programmers in that environment were low.

I talked to a number of the programmers personally, and while there were a few that I would qualify as good, most of the people on the team were average—at best. But this executive believed his people were great, and he had no interest in learning otherwise. This meant that his projects were destined for mediocrity at best and failure in many cases.

It's irrelevant what you *think* the quality of your employees is. What matters is what the level of quality truly is.

WAKE UP

So why do companies not realize how critical the quality of their employees is? The issue is complicated by a couple of factors. First, there is no single set of values used to measure the productivity of two companies. Rather, it is the combination of a number of groups, including new-product design, manufacturing, quality control, distribution, sales, marketing, etc., that each contribute to success or failure.

Because all these functions are each handled by different teams within a company, there is no single team and piece of the competition to point to. Every team can blame every other team in the company for its losses.

Second, no one views the problem as the quality of the employees. Maybe the line manager understands this, at least partially, but as you go up the food chain in almost any company, employees are viewed as replaceable. The problem is always perceived as being with management, procedures, or outside events.

This is not to say the quality of the employees is the only criterion that matters. Bad management can totally screw up even the best teams. Bureaucratic procedures can immobilize teams full of self-starters. Outside events, like the Internet, can sideswipe the best laid of plans.

But the bottom line is that without excellent employees, a company cannot thrive. And while this is the most common source of problems, companies do not recognize this as being the problem. Instead they try to find solutions by hiring more ineffective managers and piling on more straitjacket procedures, thus making the problem worse.

HIRING AT MICROSOFT

Now let's look at Microsoft. It consistently attempts to hire people with "smarts"—people who are in the smartest 5 percent. Smartest how? Well that depends on the job. A product manager is very different from a programmer. And both are a lot different from an attorney or accountant. It's *smartest for that job*.

And the key word here is *smarts*, not *knowledge*. People can acquire knowledge pretty easily (although don't hire too far afield—a programmer will probably never become a doctor).

But smarts, the ability to turn on the brain and think, are much harder to find and virtually impossible to teach. Most people are born with a reasonable amount of smarts and a lot of curiosity—the two key elements. The trick is to find those who made it through the school system without these critical elements being beaten out of them.

This is very different from the oil company practice of hiring only candidates with a 3.5 or above GPA. Grade-point average does not measure smarts or initiative, and those are the critical pieces you need.

So Microsoft says it only wants the best. Does it mean it? Absolutely! Nathan Myhrvold announced about a year ago that Microsoft "had all the budget in the world" to triple its R&D budget (amounting to an additional $4 billion); however, the "problem was finding the right people."* Rather than hire from the second 5 percent, they chose to limit the expansion of R&D.

At Microsoft I was interviewing a candidate with a 3.8 GPA from Harvard. His knowledge and programming competence were beyond reproach. Yet he could not think.

He could solve numerous standard programming problems well, but he was lost on questions outside his area of expertise, questions that measured smarts rather than applying knowledge.

Every person who interviewed him gave a thumbs down, for the same reason: an excellent programmer but he couldn't think.

---

* Eric Knee, "One on One: Nathan Myhrvold," *Upside Magazine;* March 20, 1997.

(One of the few ways to impress Bill Gates is to outthink him on an issue. Only if you can do this will Bill respect your opinion. Like every other company you have to suck up to rise. But at Microsoft you also have to be really smart, because only if you're smart can you successfully suck up.)

The employees at Microsoft doing the hiring are bound and determined to hire only really good people, because they are going to have to work with them and no one wants to have to carry someone.

> *One of the enjoyable things about working at Microsoft is that even the least talented are pretty damn smart, the average are superb, and the best leave you trying every day to match their work.*

THE MICROSOFT PROCESS

All this brings us to the interview process at Microsoft. While human resources finds most of the candidates (employees find a fair percentage), all interviews and the final hiring decision are done by the group doing the hiring.

Four or five people in the group that is hiring for a position will interview the candidate. Usually these are potential coworkers, although it is not uncommon for employees that would report to the candidate to interview the candidate too. Each person has a one-hour, one-on-one interview with the candidate.

These interviews are harrowing. For technical interviews, there is generally programming question after programming question where the candidate has to write short programs to answer each question. For program manager positions, applicants are given situations and are asked for specifics on how they would resolve them. If the answers are in generalities, the candidate is pushed for specifics. (General answers usually indicate that the candidate doesn't know the subject well.)

Each interviewer, as soon as the interview is over, sends e-mail to all other interviewers, starting with the words *HIRE* or *NO HIRE*. No equivocating allowed. The root question is: Do you think this person should be hired?

The questions also search to see if the candidate can think. One of the old favorites at Microsoft was to ask how many gas stations there are in the United States.

The interesting part was not the final answer, but how people went about solving it. Some candidates would give this question only 30 seconds of thought and then say they didn't know. No further thought or attempt.

At that time the interview was effectively over. It didn't matter what else they could do. If they were incapable of turning on the brain, they were not good enough.

People interviewing later in the afternoon, based on the earlier e-mails, will refine their questions to drill in on areas where the earlier interviewers thought the candidate was weak. The entire purpose of the interviews is to push the candidates in every direction until they fail. By doing this it is possible to learn exactly what they are and are not capable of contributing.

If the reviews are favorable overall, then the candidate is also interviewed by the prospective manager at the end of the day. The manager will again put the candidate through the wringer for an hour or so.

The manager then makes the hire/no hire decision based on his or her own impressions and the comments from the other people who interviewed the person. The human resources department is not a part of this decision. So a new employee who comes on board is joining a group that wants that person.

WHAT NOT TO DO

What is of special note is what is *not* done under this process:

First, you never have the situation where the human resources department determines whom to hire and the team first meets the person when he or she shows up for work the first day. (Note to people in the hi-tech industry: Yes, lots of companies actually do this—stupid huh?)

Second, you do not have situations where you find out, after a new employee starts, that the employee embellished his or her résumé to the point of fiction and is basically incapable of doing the job or doing it well. The interviews are direct, many times impolitely so, maybe even to the point of rudeness. But at the end the team has learned what the candidate can, and more importantly, cannot, do.

Third, a degree is not required. In software you have numerous brilliant people who have not even graduated from high school, and the industry is used to paying them just as well as people with degrees. There is no reason in any industry not to hire a qualified person merely because that person does not have a degree. All that matters is how well the person can do the work.

Fourth, this system does not guarantee that all qualified people are hired. It accepts that many qualified people are turned down. In return, it ensures that it is very rare that an unqualified person is hired. This is the basic trade-off. Either you can hire everyone who is qualified and many who are not, or you can turn down some you shouldn't and very rarely hire someone who is unqualified. But only with this careful approach do you end up with a superbly qualified team.

> *For every hire keep in mind this critical corollary: The new hire should raise the mean of the team. If the team is all in the top 5 percent, then to raise the mean you are really trying to hire from the top 2.5 percent.*

OK, so this makes sense for an intellectual kind of company like Microsoft. But what about other companies? The same approach holds. But it's critical to make sure you find the correct way to measure employees. After all, a programmer would be terrible working at a fast-food counter (no social skills), as a lawyer (won't wear a suit), on the factory floor (will never show up on time), or as a carpenter (not in the greatest shape).

If they are really good, workers in any type of company—a fast-food restaurant, a bank, or a factory—are more concerned than management about making sure that new hires will do their job properly. And if properly trained, workers can do a better job of making that determination because they are closest to the job.

Smarts are critically important because smart employees are the ones that will catch errors sooner.

And they will come up with a more efficient way of doing something, thus saving themselves work and saving the company time and money.

OUTSOURCING

An employee needs to be more than smart. He or she needs to be committed to the common goal. Outsourcing is killing this sense of commitment (at Microsoft as well as at other companies).

This practice saves companies thousands of dollars in personnel costs—and costs millions in lost productivity and reduced quality. First, it's a basic violation of the "hire the best" principle. After all, if someone is in the top 5 percent (or even the top 20 percent), would the person be working through a temp agency?

Second, the end result of outsourcing is that contractors are treated as second-class citizens (i.e., no benefits, etc.) so they cannot legally argue that an employer-employee relationship exists between them. This virtually guarantees morale problems, which in turn produces unmotivated workers who do not feel they have a stake in the common goal. It also guarantees a higher turnover rate than among the "first class" employees. All of this leads to a loss in efficiency and a loss of quality that, while much harder to measure than the bottom-line cost savings, is very, very real. Don't outsource any key function; otherwise, you are putting your success in the hands of individuals with no real commitment to your success. And without that level of commitment in your work force, you cannot win.

In another 10 to 20 years outsourcing will probably rank right up there with New Coke as one of the biggest business mistakes of all time. Only this one is devastating company after company.

FINAL WORDS

You are only as good as the people that report to you. If you do not insist on hiring absolutely the best, then you are consciously deciding that employees of average or worse quality are acceptable. The ground rules have changed. You can no longer safely assume that your competitors will hire at random. For the first time you will be (or are) competing against companies that do hire the best.

Hiring the right person is *the* most critical decision a manager makes. Determine the criteria for the perfect employee and hire only people who fall into the top 5 percent of that category.

No matter how desperate you are to fill a position, no matter what the consequences, *do not compromise*. Spend the time necessary to make the right selection—the quality of your employees determines if you can succeed. It's that simple.

3

BET THE COMPANY

Every year at the company meeting Bill would say the same thing. He would repeat it again at periodic strategy meetings. We all heard it so often that it basically went in one ear and out the other.

"We are betting the company on Windows." Three years later, "We are betting the company on the Internet." OK, fine, we know we're betting the company. Tell us something we don't know.

Very few companies are willing to bet their future unless they have no choice. Even though Microsoft doesn't need to bet the company, it chooses to do so, and truly does do so, year in and year out. And doing so is taken for granted.

BURNING THE BOATS

> *The Viking boats landed, and the Vikings roared ashore acting in a manner akin to Microsoft's actions in the software industry. They thundered across the beach, captured the town, and seized the fort on top of the hill. They then looked back down on the beach to see their boats being burned to the ground on the orders of their captain. Why burn the boats?*

You burn the boats to say that you are staying forever. Returning is not an option, even if things go badly. No matter what, the only course is to move forward. Microsoft burns its boats after every victory. Defeat is not an option.

KILLING THE CASH COW

Microsoft used to be called the house that DOS built. MS-DOS was an incredible cash cow, generating the majority of Microsoft's revenue for all of the early years.

MS-DOS held 80 to 90 percent of the PC market. It sold with every new computer. It required virtually no advertising and minimal development costs. It was basically pure profit.

MS-DOS is no longer being sold. So who took this incredible profit stream away from Microsoft? Who killed Microsoft's bread and butter? Microsoft did—intentionally.

As soon as a Microsoft product owns a market, Microsoft starts looking at how to eliminate it with a better product. Microsoft killed DOS with Windows. It is trying to replace Windows with Internet Explorer. Word, Excel, and other products are all constantly being improved.

Microsoft tries to kill its own products for two very important strategic reasons. First, if it doesn't do it, someone else will. It is very aware that every time there has been a major paradigm shift in the software industry, every major company except Microsoft has lost its position in the shift.

The companies leading a paradigm shift within a market are the ones that will dominate in the new market. Paradigm shifts are also opportunities to gain new markets. The large, established players that wait out the shift and then try to change are destined to fail, because they cannot change fast enough after the fact.

The second reason Microsoft tries to kill its own products is to force a paradigm shift. This creates confusion and in confusion there is opportunity. Remember, in every paradigm shift the established software companies have either gone out of business or become much smaller.

And into this opportunity steps . . . Microsoft. The shift from DOS to Windows allowed Microsoft to take the word processing and spreadsheet markets. The shift from mainframes to client-servers allowed Microsoft to take the low-end database market.

But the bottom line for any company—especially over the next several years as the Internet brings a major paradigm shift to almost every industry—is you can either lead the change or be collateral damage as the change rolls over you.

CHANGING COURSE

One day the Internet appeared on the radar. The Internet had existed for 20 years, and the Web had been around for awhile by this time. Microsoft used the Internet—the company has always lived on e-mail—so the Internet was not new or unknown. But suddenly the Web was starting to explode. Non-techies were using it. Netscape appeared on the scene. And the Internet threatened Microsoft. It threatened its core monopoly (Windows), it threatened its strategic ownership of the desktop, and it threatened the computing model at the core of Microsoft's business by moving much of the computing from the desktop to the Web server.

How would most companies react to this? By vigorously defending the status quo. We see this today in the telecommunications industry as the Baby Bells try to slow down change by manipulating the regulatory process and steadfastly refusing to compete against each other.

To many people, defending the status quo would have been the smart thing for Microsoft to do. After all, Microsoft held no advantage in the Internet game, and it was actually behind Netscape and others.

At Microsoft Bill looked at the Internet, saw what was coming, and embraced it. The marching orders went out: The future of the company was now the Internet. This bears repeating. Microsoft chose to bet the company on a new paradigm in which it was starting out at a disadvantage. And it did this to the detriment of its existing products.

And what about all the strategic plans for the next several years? The designs, projects, marketing campaigns, etc.? All canceled wherever they conflicted with the new direction. The past plans and direction were thrown out with nary a second glance. So what was the result? In nine months Microsoft went from having no Internet strategy to being an Internet-focused company. And at the time of this writing, it's pretty much over for Netscape.

Could any other Fortune 500 company change course 180° in nine months? Or even in nine years? Because Microsoft could, it was not only able to compete with Netscape, but is well on the way to dominating the Internet after a late start.

ANZIO

During World War II, the U.S. Army, under General Clark, was fighting its way slowly up the Italian peninsula. The Germans, under General Kesselring, were doing an awesome job of making the American advance as slow and expensive as possible.

The Americans decided on an amphibious attack north of the German line at Anzio. The German forces would then be trapped between the two American forces and would have to surrender. Once completed, all of Italy could then be easily captured.

The American forces, under the direct command of General Lucas, landed on January 1, 1944. The Germans did not know about the landing until after it was almost completed, and everything rolled ashore smoothly.

The Americans did not know where the Germans were and sent out several reconnaissance patrols. One jeep with a sergeant and two privates drove from Anzio to Rome and back seeing no Germans. The path was clear to take both Rome and the German forces from behind.

However, General Lucas was a careful man. He figured the Germans had to be around. He was not willing to commit his forces until he found the German forces. So he dug in and waited. And the Germans swung around and attacked. The troops at Anzio were kept bottled up until June 3, 1944, when the main advancing American line was able to push that far north.

Because General Lucas was not willing to bet his forces, he incurred enormous unnecessary casualties while not bringing the end of the war even one day sooner.

If you're not willing to make bets at good odds, then you can't win.

STOP

Several years ago Bill, Mike Maples, and others realized that growth was out of control at Microsoft. The staffing was increasing at 40 percent per year and the controls in place were not sufficient to manage the number of people Microsoft now had.

In addition, because Microsoft made money in almost every market it went into and because it had such incredible profits, there was no financial brake on the growth potential of the company. In fact, the only real limit on growth was available office space.

I'll bet at this point most people are reading this and saying, "So, what's the problem?" Well, the problem was that Microsoft was on the verge of becoming the typical large corporation, with each division operating as its own independent fiefdom. And with Microsoft's cash reserves and profits, it would have had the luxury of getting itself really screwed up before financial problems forced it to fix things.

Microsoft reacted by putting an absolute lid on hiring. It went from increasing staffing at a rate of 42 percent per year to under 5 percent per year. It was wrenching. Numerous projects were adversely impacted because the freeze hit everyone.

And over a year's time, changes were made in how the company was managed, products in nonstrategic markets were dropped, and effective control was regained. This was done because senior management understood that effective control was more critical than immediate growth and profits. And only because this control was regained was Microsoft able to respond to the Internet.

APPLY IT

Those who equivocate, lose. You cannot win against a competitor who is totally, 100 percent focused on a market if you have part of your organization covering other options. When there is a paradigm shift, embrace it. The winners will be those companies that use the paradigm shift to their best advantage. The losers will be those companies that fight to maintain the old way of doing business.

You must be nimble and be able to react instantly to changes in the market. Things are changing faster and faster every day and there's no slowing in sight. In fact, change will come even faster.

What's the lesson here? Make your own products obsolete. For most industries, both products and services sold today will almost certainly be obsolete in 10 years, and possibly even 5. If you wait for someone else to replace your product first, then you've given away that market.

Security is no longer an option. If you don't bet

the company, if you don't fully exploit the paradigm shifts, then your competitor who does will take your market. Not making the bet is no longer the safe move. Instead it is the one move that guarantees that you will fail.

4

REQUIRE FAILURE

S NAFU originated, I believe, in the Army. It stands for Situation Normal, All Fouled Up. (*Fouled* is the polite substitute for the more common F-word in the slogan.) At first glance it seems like a pessimistic statement, that the norm is a fouled-up operation. But it also speaks to the disorganization and miscommunication inherent in any organization made up of people—especially an organization that is constantly having to adapt.

The trick is to understand that a company that is capable of responding effectively to rapidly changing market conditions will operate in SNAFU mode. It will aspire to do better, and many times will. But if it tries to force a better operating mode at all times, it is going to either be slowed down to ineffectiveness or be constantly disappointed.

FAILURE IS A REQUIREMENT

How do bugs get into computer programs? It's very simple. Every time programmers sit down to write part of the program, they put them in there. Think about it—the only way to get bugs in a program is for programmers to put them there.

Note that programmers do not do this on purpose. These bugs are due to conditions they did not consider, incomplete understanding of the problem or system, mistakes made while typing, etc. But they are typed in by the programmer. Because human imperfection leads to numerous bugs, a software company must accept the fact that its employees are imperfect, that failures are a constant part of creating programs.

But does this hold for other types of business? Yes! It's just not as obvious because it's not as severe a problem. An employee at McDonalds' can go through an entire shift making no mistakes. But it is not reasonable to assume that at no McDonalds' will any mistakes be made.

Or FedEx can commit to deliver every package overnight and have the vast majority of its employees execute delivery flawlessly. But at the same time, it has to accept that some packages will not be delivered in the promised one day. And in some cases (like some packages I have sent), it can take a week or more.

Failure is expected. And therefore reasonable failures should not be a reason to reprimand people (except when FedEx is late with my packages!).

SUCCESS IS THE REAL METRIC

Garry Marshall in his book *Wake Me When It's Funny**(good book) said that the life of a network executive is quite simple. If the executive approves a show and then it fails, the executive is fired. But if that person approves no shows, his or her job is not at risk.

Most large businesses work this way. To succeed is good, but to fail is unacceptable. Think about it. What happens if you fail at your company? Odds are, it's a major problem for your career.

Now what happens if you succeed? In most cases, it doesn't matter much. In fact, the real question usually is: Did you succeed without stepping on too many toes? Because success at most companies, with rare exceptions, does not excuse anything.

* Garry Marshall, *Wake Me When It's Funny;* Newmarket Press, New York, 1997.

At Microsoft it is the exact opposite. Failures are expected, and as long as they are not exceedingly stupid, they are basically forgotten.

FAIL QUICKLY

The people who design jet engines use a chicken test. This test fires chickens (usually purchased at the supermarket) at a running engine. They attempt to run this test as early in the design process as possible because if the engine can't pass this test, there is no point in spending additional millions designing it.

> *This story may be an urban legend but it's a good one: A British company asked Boeing for one of its chicken guns to test a new jet windshield. After using it the Brits called up Boeing and reported that the chicken went through not only the windshield but also the brick wall behind it.*
>
> *Boeing sent an engineer over to England to investigate. After watching the workers run the test again, Boeing added to the instructions, "Make sure chickens are defrosted before firing."*

The key is to identify failure as quickly as possible. Sit down and try to come up with everything that could lead to failure. It's often easy to accurately predict what can trip you up. The surprise is usually in which of the predicted items actually did cause the failure.

Then, for each item, figure out how to determine, as early as possible, if this is a showstopper. And if it is, then *don't give up.* Find a way around the problem. Only if the problem is truly unsolvable do you kill the project.

At the other end, in all too many companies, failure is so unacceptable that projects are continued long after everyone on the project knows it is destined to fail. Why? Because management has made it clear that failure is not acceptable.

At one company I worked for, I determined that one of the projects there had made no progress for the last year and was, as presently structured, destined to continue making no progress.

I laid out the problem to the president, who did not believe me and clearly did not want to hear anything negative. I insisted that a real problem existed and repeated that the project was not going to move forward in its present state. Annoyed, he finally talked directly to the people on the project, who confirmed what I had said.

The president's reaction? He was angry and felt that the people on the project had misled him by not telling him sooner. But when I originally delivered the news, he made it abundantly clear that he did not want to hear anything about it.

FAILURE MEANS SUCCESS

Fast failure is acceptable; slow failure is not. But even more unacceptable is no failure. If people never fail, then they are not trying hard enough. They are not pushing the envelope.

Failure means try something else. In most cases you can find another approach, another system, another solution that will work. Failing quickly usually means finding a successful alternative quickly, not closing down the project entirely.

Most projects that fail do not fail because there was no way to make them successful. They fail because they went down the wrong path and no one was willing to change direction once it became apparent that that path was destined for failure.

Every presentation on the status of a project to management should include what the major risk factors to the project are and how quickly the actual risk will be measured. No one in management should accept a status report without a list of the major, known risks.

At Microsoft failure is expected. No one who has been successful there has not had some spectacular failures. And in some cases, people responsible for spectacular failures have been promoted because of what they have presumably learned from those failures. (There are avoidable failures—sometimes the result of incompetence or stupidity. It is critical that you understand why each failure occurred.)

Think what this means at Microsoft. There is no penalty for understandable failures on the road to success (aside from exceedingly stupid things), and there are substantial rewards for success. So employees at Microsoft will make attempt after attempt for success without worrying about the failures the unsuccessful attempts led to.

Almost as important, the employee will not waste a lot of time on unproductive tasks meant to cover any blame for mistakes. If no one cares that you failed, then there is no need to build a time-consuming paper (or electronic) trail.

In this environment, while you will experience a lot more failures, you will also have substantially more successes. And the additional successes more than pay for the additional failures—many times over.

If failure is not an expected part of the path to promotion within a company, then the successes will be a lot fewer.

POSTMORTEMS

Immediately after each project is completed at Microsoft (and many other hi-tech companies), a "postmortem" is held. This is a process (mainly several meetings) where the project team discusses everything that went wrong and everything that could have been done better.

Note that there is no discussion about what went right, unless it is a process that may not have been apparent to everyone. The whole purpose is to figure out how to do the same project better the next time around.

All participants in the meetings are expected to be ruthless, not only on problems others had, but especially on problems they experienced. They are expected to outline what they did wrong and how they could do it better next time.

Because the judgments are minimal, people are willing to discuss the mistakes they made. In fact, there is even some pride taken in having had a spectacular failure or two—especially if you then came up with the solution yourself.

The postmortem is the leading example of the acceptability of failure and the desire for unceasing improvement. Failure is a lesson to learn from; and, further, it is expected, virtually required, that everyone make mistakes.

DELIVER BAD NEWS IMMEDIATELY

A critical part of failing quickly is delivering bad news as early as possible. If there are numerous, ongoing problems throughout a project and senior managers aren't kept fully apprised with an accurate and complete status, they tend to get very nervous.

Problems need to be reported immediately. In fact, you should usually know that something *may* become a problem before it actually does. At that time, you should be advised that a certain problem may exist, what is being done to research it, and when you will know if it is in fact a serious problem. (I personally feel that I have slipped up if the first time I tell my senior manager about a problem is when I know for sure it's an issue, because if I had been doing my job right, I would have seen it coming.) Further, just reporting the problem is not acceptable. A solution, or solutions, must be presented at the same time.

All of this requires that managers accept problems and failure as part of the process. If managers get upset every time they are given bad news, then bad news is not going to be delivered.

People need to be thanked for delivering warnings and bad news at each step in the management chain. If at any level this is not done, then that level and above will no longer be told bad news and will no longer know what is really going on below them.

At times people will get upset when you tell them about a mistake you made. It's human nature, and the worse the mistake, the more likely this is to happen.

When it does happen there are three little words that can totally defuse the situation. Those three words are:

"I screwed up."

I have used them when I have made a major dumb mistake and managers above me are ready to rake me over the coals. And it has totally deflated them. Why skewer me if I already know I blew it?

> *And when failure is expected, then management is primarily concerned that people learn from their mistakes. They also understand that occasionally someone will do something really stupid. After all, we're all human beings.*

OBVIOUS SHORTCOMINGS

I was once a member of a task force for a local school district (not Boulder!). The administration was showing me some of the software they were using and asked me what I thought of it. I replied, "It sucks."

Oh boy, did it hit the fan then. We were discussing $270 million in bond money, and nothing caused more of a fuss than using that phrase to discuss some software. And the fuss was not over the existing problem, but over the words (actually *word*) I chose to use.

On the flip side, at Microsoft we described proposed designs with words a lot worse than *sucks*.

Take a look at your own company. If you say something sucks (or use other descriptive terms), does the ensuing discussion focus on the problem or item at hand or does it focus on the words used? If something sucks, the important thing is to fix the problem. Not only is the acceptableness of the words used irrelevant, but in some cases words like *suck* are the best description of the problem.

PUSH DECISION MAKING DOWN

This is a mantra in so many management books that there is no point in going into the details again here. Needless to say, decision making is pushed way down at Microsoft.

In fact, most of the decisions made by upper management are issues brought to them by people below them who feel that because the question impacts so many other areas, someone higher up in the food chain needs to make the decision.

But it's more than "allowing" employees to make some decisions. It's putting the power in the hands of the employees. When a manager and an employee disagree at Microsoft, it's not predetermined who will win the disagreement.

The managers there realize that they must have buy-off from the people doing the work. Otherwise, there is no way the work will be done well or quickly (or even at all). Further, both the manager and employee know that upper management is concerned with determining what the best decision is and in keeping the employee motivated.

An on-line gaming company was bought by a large telecommunications company. The telco company brought in its own upper management to run the gaming company.

One day one of the telco managers told a manager below him that she and her staff needed to change a large part of the program. The lower-level manager said she would ask the programmer involved if he would make the change.

The telco manager blew up at the concept of having to ask an employee if he would change his program. He was an employee and he damn well better do what he was told!

It's no great surprise that the telco company ended up selling the gaming company at a substantial loss.

At Microsoft employees have no worries about disagreeing with a manager. The simple act of disagreeing, with valid reasons, is expected, and everyone does it. Therefore, it does not adversely impact one's career. (At least not in any obvious way. At times one person will get royally upset over another's critique, but to come down directly on someone for it would be unacceptable.)

Because employees at Microsoft believe in what they are working on, they will fight for what makes sense. This willingness to fight and the lack of serious consequences for disagreeing with management give the employees phenomenal power.

There is one final note to all of this: When disagreeing, an employee is expected to have concrete, valid reasons supporting his or her viewpoint and must provide a superior alternative. Stating that you just don't like something is not acceptable.

FAST DECISIONS

Because employees have the power to make almost any decision relating to the work they are responsible for, decisions can be made, and are made, quickly. The only reason to put off making a decision is if the employees themselves believe they should check with others first.

This desire to make a decision and move on pervades the entire company. At any meeting it is considered bad form to leave a decision pending until the next meeting.

To make this work requires that people presenting a proposed solution outline any reasonable alternatives and the pros and cons of each of the alternatives.

The first time I saw someone do this at Microsoft I couldn't believe it. The person was basically providing all the ammunition against his own argument, as well as the information in favor. But the system works well. If you can list the valid reasons against, as well as in favor of, your proposal, then you understand the problem well. And sometimes in working through all of the arguments, you decide an alternative solution makes more sense.

THE FREE FLOW OF INFORMATION

The most critical component to all of this is accurate and complete information. People can generally make good decisions if they know what is going on in associated parts of the company. If they do not know what is going on, people will almost certainly end up working at cross-purposes.

The primary means of disseminating this information at Microsoft is e-mail. Reviews, proposals, problems, designs, architecture, etc., are all e-mailed to everyone who may be interested. (If in doubt, you add someone to the CC line—they can always choose to not read it.)

People will comment by replying to the e-mail. In this manner, at each person's convenience, all who are interested can clearly state their arguments concerning the issue. Sometimes this correspondence will be followed with a meeting to make the final decision, but usually the e-mail alone is sufficient.

By sending this information out so widely, everyone impacted can comment on the proposal. Everyone involved knows what is going on, and by using e-mail, everyone has the chance to carefully list out any concerns on an issue and to be heard.

This also forces, at least at Microsoft, people to cut to the essentials. No one is going to read e-mail from a person who rambles on and on. Arguments must be persuasive, and to the point, in order to be accepted.

A senior vice president at a bureaucratic company once told me that all the status reports he received were so watered down that he did not really know what was happening within the company.

I felt sorry for him until I remembered that he had been at that same company for over 10 years. He could get the information if he really wanted it. Employees are more than happy to tell management everything that is wrong, if there are not repercussions to the disclosure.

He was at least aware that he didn't know what was going on, but he obviously had no real desire to discover all the problems.

5

MANAGERS ARE
QUALIFIED

MANAGERS KNOW THEIR INDUSTRY

I remember one time when I went to talk to a vice president at Microsoft, at my request, to discuss some major problems I saw on a project. As I outlined the various problems, he would add details for me. It soon became clear that he had a very, very clear picture of the situation—in full detail.

Now compare this with your average mid-size or large company where the executives at the top have no idea what is going on. (Nor do they have any real desire to actually learn what is happening below them.)

But there is an even more critical point to this: Managers at Microsoft fully understand the work the people who report to them do. Almost without exception, those managers could do the job of any individual doing the core work for their team.

Managers of programming teams are all programmers, and not just people who could write a program if a gun were put to their head. Almost each and every one is a good to awesome programmer—including Bill himself.

The people managing the marketing teams are marketers. The people managing the sales teams are excellent salespeople. Scott Oki, who was the vice president of sales in Microsoft's formative years, was probably the single best salesperson in the company, as well as an excellent manager.

A manager who is not capable of doing the work the people who report to him or her do is incapable of effectively managing those individuals. This holds true for a number of reasons.

First, the people doing the work will not respect a manager who can't do their job. Right or wrong (and it is right), people will not respect an uneducated opinion on how to perform their job.

Second, how can managers make decisions on

issues brought to them if they aren't capable of doing the work? What will they base the decision on? It generally ends up being based on who is the better presenter. And equally as often than not, that is the person with the weaker grasp of the problem.

Third, how can managers truly understand the status of their group if they don't understand the work? They can be blinded by smoke and mirrors almost until the day the product actually has to be delivered.

The counterargument will be that this cannot possibly be valid because it means that the majority of managers in America are unqualified for their position. Numerous business practices, practices followed by the majority of companies, have been found to be ineffective and later eliminated. The fact that everyone does it doesn't mean it's effective.

The simple fact is that Microsoft has better managers throughout its organization. They are not perfect by any means, but on average they far surpass those in most other companies. There is one simple reason: The primary qualification for managers at Microsoft is their expertise within a particular field. Management and people skills are of secondary importance. It's worth saying again: Managers are required to have very strong technical (or other appropriate professional) skills; and while other management talents are taken into account, the company accepts that those other talents may be moderate at best.

This applies all the way up the organization, even to Bill, who is an excellent programmer, an excellent marketer, and an excellent salesperson. And equally important, one of his *primary* areas of expertise is programming, which is Microsoft's core product.

The military understands this concept. Combat experience is almost always a requirement to be promoted into the upper ranks, because someone who has not been in combat cannot possibly understand the trade-offs inherent in battle.

And officers in the equivalent of a middle-management level are moved from command to command every two years. One of the major reasons is to give them experience with multiple parts of the service while they are still close enough to the ranks to see the work actually being performed.

This is why Microsoft has prospered and will continue to prosper at their competitor's expense. Other companies, including IBM, promote salespeople to the top. Microsoft promotes programmers. (OK, and Steve Ballmer too.) Microsoft will win the battle every time.

MAKING THE SWITCH

A company with qualified managers is going to dominate a company with unqualified managers. When all companies had equally unqualified managers, this didn't matter. Now that some organizations realize how key this concept is, the companies that don't are easy prey for those who do. Ignoring this issue is no longer an option.

So how do you make the switch? Well, no one ever said competing with a company like Microsoft would only require one or two easy changes. This is a major effort and much more difficult than instituting new policies.

And good, qualified managers are not lined up at your door waiting to be offered a job. (They're lined up at the door of Microsoft or other companies that have already made themselves more efficient.) Obviously, you need to get from here to there without destroying your company on the way.

As a start, take a hard look at your management staff. Many of the managers you presently have either are qualified or, with a reasonable amount of training, could be qualified. By and large the management position they hold didn't require them to be qualified, but they are. Lots of people were promoted from within the group they now manage and they do have the proper skills. You will probably also find that by and large these are your best managers.

Other managers, while not qualified for the position they presently hold, are well qualified to manage other groups that do match their expertise. A moderate amount of shuffling can solve some of your problems.

In addition, there are people currently working for you who could be promoted to the next level as a manager. And these are people who do understand what their groups are doing. Not only does this solve the management problem, but it also boosts morale because you are promoting from within and promoting people who actually understand the work being done.

For the remaining managers who are unqualified, you need to make it a priority within the company to replace these people as quickly as possible in a manner that does not have those about to be replaced leaving before suitable replacements are found.

NOT A MODELING AGENCY

Aside from a few select job categories, such as actors and models, there is absolutely no benefit in requiring managers to dress nice, look nice, or have good hair. Once again, how managers look has no impact on the quality of the job they do.

When promotions are limited to those who look the part, who came from central casting with the "manager" look, you are drastically limiting the number of candidates you will choose from and thereby significantly reducing the odds of finding the best candidate.

DOG EAT DOG

While not a perfect meritocracy, promotions at Microsoft are always almost based on ability. At the same time, Microsoft is expanding extremely fast, and existing departments seem to be reorganized every nine months.

The result of all this change is that there are constant opportunities for promotion at Microsoft. And these opportunities go not to the person who has been waiting 10 years and is next in line, but instead to the best person available.

This means that there is constant jockeying for position at Microsoft. There is always another position opening up, and the best candidate is the one promoted. As a result, there is a constant battle among many managers who are each attempting to get that next promotion.

The controlling factor is that people are expected to perform, and once someone gets a position, he or she needs to produce. However, as managers move up the ladder, the competition gets more and more cutthroat (very much like being on a Klingon warship).

This is not the most pleasant work environment ever encountered, but it is very Darwinian in that it effectively promotes the more capable personnel at the expense of the less capable. Effective managers are moved up. Ineffective managers are moved down or go on permanent sabbatical.

And because promotions are based primarily on performance, there is a direct reward for doing a good job. This selection process keeps people focused on doing the best job they can, both to improve their odds for promotion and to protect their existing job. What about those who don't like the competition? They generally max out several levels below where they would be in a less competitive environment.

This ensures that the higher you go in the management food chain, the more you will find extremely competitive managers who create very productive teams. And for the company, that is much more valuable than a happier and calmer, but less productive, environment.

Microsoft managers fight extremely hard for each external market because it's the same thing they do to get promoted internally. And in both cases they deliver 100 percent all of the time. Managers at an old-line company, where it is critical to get along to get promoted, can never successfully compete against this. In fact, they will never even understand what they are competing against.

6

PERFORM, PERFORM, PERFORM

WHAT ARE YOU GOING TO DO FOR ME TOMORROW?

OK, so you created a new business unit, entered a new market segment, and dominated it. So what? That was yesterday. What are you going to do for me today?

At many companies a big success can carry you for the next one to ten years. At Microsoft it is an indicator of how successful you may be on your next job, but little more.

Microsoft gives you virtually nothing for what you have already done. You never get to coast on previous work. And, in fact, if you burn out doing an awesome job, you are cast aside once the work is done because you are no longer of use to the company.

This is a very scary concept for most employees—and most companies. It means that you can never stop and rest. If you do, someone else who is not resting will pass you by and you may become expendable.

It also means the corporation is totally heartless. Ten years of dedicated service means nothing. Every day you have to continue proving your worth. (On the flip side, its success has made it unnecessary for Microsoft to lay off divisions, including people who are doing a good job for the company—a common occurrence in all too many companies today.)

The result is that all the employees are constantly challenged to do the best job they can for the company because the company is constantly looking at what the employees are doing—now.

PERFORMANCE, NOT EXCUSES

No one at Microsoft is much into excuses. If there are problems, people want to hear the solutions. Everyone is focused on the goals, and excuses just slow down problem solving.

When success is the only measure, then excuses become irrelevant. And at Microsoft success is what people measure things by. Everyone realizes that some problems are harder than others, and everyone realizes that bad luck occurs. But it is also irrelevant.

If a product is not being shipped, then all the excuses in the world won't ship it. If a marketing campaign is failing, then all the excuses in the world won't fix it. The only critical question is: What is going to be done to fix the problem?

Some people will claim that this is unfair. Yes, it is. But Microsoft has never been concerned with fairness. It is concerned solely with success. An extremely successful environment does not have to be fair.

And by the same token, Microsoft does not need to be fair to attract highly motivated employees. Many people are highly motivated in a success-oriented environment.

Let's take a look at a sports team. Pro or city league, it doesn't matter. Everyone understands the rules; you win or lose. If a player sprains an ankle, the team doesn't get a couple of extra points. If a player is having a bad day, the other players generally aren't terribly understanding.

Are the fans understanding of excuses? They don't want to hear it. Maybe for a death in the family the player will be cut some slack for a game or two, but that seems to be about it. All that matters is winning.

Let's look at it the other way around. Assume each player on Brazil's team in the '98 World Cup had a good reason to be off. One was breaking up with his wife. Another was worried about his kids in school. A couple had sprained muscles that really hurt. And one got no sleep the night before because he was so nervous about the game.

So are we all now understanding? Is it OK that they played the game the way they did? Do we all feel like it is no big deal? No. Because what everyone focuses on is that they played terribly. Excuses are irrelevant, and virtually everyone is very comfortable with that viewpoint.

Does this give us an unmotivated team? Actually, it gives us the exact opposite; a highly motivated team. Imagine the result if a sports team accepted excuses. Performance and morale would both plummet.

Microsoft attracts employees because it is success-oriented. Both morale and performance are stellar due in part to this focus. People want to work with successful groups, and they are willing to accept that they are judged on their success alone as long as everyone else is too.

In fact, the people who do not find this environment acceptable, if not preferable, are generally people who are not comfortable being measured on their success. And this discomfort generally comes because they are not terribly successful.

So this success-oriented environment self-selects employees. The most successful are generally those promoted, and the least successful are generally those who leave or are never hired.

Two teams of people were working on a project. One team was from a large company. The other was from a small company that had been acquired by the large company.

The project was slipping. There was a phone call between several people on each team. The purpose was to figure out how to complete the project on time.

The team from the small acquired company was focused on what needed to be done and how it could be accomplished as quickly as possible.

The team from the large company insisted on spending three days documenting why they couldn't complete the task on time. Documenting the excuses was more important than being successful.

RATING EMPLOYEES

So how are employees rated? Again, on their successes and failures. For something major that was out of the control of the employee, some weight may be given to a reason for failure. But this is rare. By and large it's performance that matters. And because it's performance that matters, that is what people concentrate on. If no one will pay any attention to excuses, then no one will put much effort into formulating them and will instead concentrate on performance.

This can have a nasty side effect. I have seen several cases where it was "common knowledge" that someone had screwed up real bad, in either a bad design or a bad piece of coding. Yet in these cases, "common knowledge" was dead wrong. The individuals in question had done it right. However, because there is no mechanism for discussing excuses, those people were unfairly left with the failure on their reputations.

So failure is expected, and yet it may count against an employee in his or her performance rating. How can this be? Microsoft's core business, software development, marketing, etc., involves a process in which the road to success is littered with failures. On average, about 50 percent of an employee's time is spent on things that the employee later discovers won't work. So in order to succeed, you have to fail. And it is the end result that is measured by its success or failure. While the failures along the road to completion are by and large both expected and ignored, the rating of the employee is based on what was accomplished. If the project itself is a failure, then that will affect the employee's rating negatively. However, because failure is also expected a certain percentage of the time, the failure is not a permanent black mark on the employee's record. It is just a measure of the work the employee most recently completed.

In other words, failure is cause for a negative rating. But it is also expected to occur for each employee occasionally and therefore does not have any long-term effect unless the employee continues to fail and fail and fail.

A more global form of this is projects that failed for market reasons. The initial Windows multimedia effort was staffed with a number of very good people. They created a good product, but the market wasn't ready and the product failed. As a result, everyone on the project was marked with a failure.

No system is perfect. While some can play the success game better than others, jumping to different projects and being a source of the "common knowledge" rather than the brunt of it, Microsoft's system generally treats everyone pretty fairly. Further, people are not fired because the project they were on failed. Rather, they are looked at for their individual work and judged on how successful they were within the scope of the project. And even more critical, while the system will at times treat individuals unfairly, it works very well for the company. It keeps everyone focused on performance, not excuses.

This is one of the major reasons that appearance is so unimportant at Microsoft. How you look, how you carry yourself, how you talk—those are all secondary to how productive and successful you are.

This is not to say that other factors are completely irrelevant. Here comes the disclaimer: Even Microsoft is not perfect—surprise, surprise. People are people and even at Microsoft looks, personality, gender and skin color affect individuals' perceptions of others. It's worth noting that Microsoft's top management team is, and has always been, made up primarily of white men. The difference is that performance is given significant priority over these other items, whereas at most companies performance and success seem to be secondary at best.

When performance and success are how you get ahead, people concentrate on being successful. People are not stupid. They respond very well to what truly matters to a company. (And they are also very good at determining what truly matters, as opposed to what the company says matters.)

PUSHING THE ENVELOPE

MIT starts off its freshmen by pointing out one simple fact: Virtually each and every one of them was easily the smartest student at his or her high school. Far above every other.

The students are then invited to look at the others in the auditorium. These are the students who were the smartest in their high schools. Not only are those others as smart; many are smarter —much smarter.

I felt like a moron when I first started at Microsoft. I had come from an environment where I was one of the best, and suddenly I was surrounded by people who were as smart as me. And some who were a lot smarter.

However, even more key was that at Microsoft,

they were used to working with people as smart as they were. And so they had learned to think things through and to really understand both sides of an issue before opening their mouths.

For the first six months I scrambled like crazy to catch up. I was personally determined to become as good as I could, and because of this competition, I became a much better programmer than I ever would have become without it.

This is not the dog-eat-dog, only-one-person-wins kind of competition. This was the fact that I believed I was not as good as my peers and I was determined to be as good as they were, both for my own self-respect and for the sake of carrying my fair share of the weight for the team.

In a highly competitive, success-oriented environment such as Microsoft, how do you succeed? Not how does the company succeed, but how do you personally stand out? Knocking yourself out doing a damn good job won't cut it. Everyone around you is doing at least that. And every one of those people around you is at least as good as you, or better. In a competitive environment like this, you want to succeed because success is how everyone is measured.

The only way to do this is to push the envelope of what you can do. Every day try to do better. Work smarter. Work harder. Innovate more. People are focused 100 percent on performing their job as successfully as possible.

This is not to say that people are looking over their shoulders at their coworkers all the time. Most of the time at Microsoft is spent working—getting the job done and doing it well. And a lot of self-satisfaction comes from doing that. But the satisfaction is two-fold, both the joy of a job well done and the knowledge that this is what is measured.

It's when choosing which project to do next or choosing how to implement a job assigned that this comes into play. The choice is almost never to do as little as is required; instead the choice is to do as spectacular a job as possible.

You therefore have virtually every employee working as hard as he or she can to do the best job possible for the company. And because this provides personal satisfaction, as well as corporate advancement, employees receive double the positive reinforcement for their efforts.

SHIP IT!

When Windows 3.0 was nearing completion, signs sprung up on doors everywhere in the Win 3.0 group saying "Ship It!" What everyone meant was that the product was good enough, so get it out the door and into customers' hands. Now granted, Win 3.0 was (very) far from perfect, but perfection takes forever, especially for software.

What everyone was focused on was that when the product was good enough, it was time to ship it. And everyone understood that the job was not finished until the product was shipped. Having completed your part didn't matter if the rest was not ready, so virtually everyone, except those few scrambling to finish their parts, wanted it shipped now. Because that was the measure of success.

This permeates every project I have ever seen at Microsoft. No one has ever said that their part was done so they've been successful. No one has ever considered their performance excellent if the project is in trouble. People take ownership in the entire project.

This feeling of ownership has a couple of very important ramifications. First, because people take ownership, they will step in and voluntarily help where needed. Since their job isn't done until the project is shipped, people with less work feel compelled to assist others with too much.

Second, people pay a lot more attention to the entire project. If another part is in trouble, the sooner it is addressed, the sooner the project is back on track.

I have rarely heard people say another part of a project is not their concern. And this is not just managers; this is everyone.

Third, managers have to keep everyone on the project informed of direction and status. And for major changes in direction, they have to get buy-off from the employees. This is a critical component. People will not fight and die for imposed goals, but if they helped set the goals, then they will.

Employees are not stupid, at least not the ones at Microsoft. They won't push to do something that won't sell. In fact, many of the questions asked when goals are presented bring up valid issues and many times lead to changes in those goals.

Does this mean that everyone agrees with everything? Absolutely not. Some will think the basic direction is flat-out wrong. But the critical point is that everyone was heard, their concerns were addressed, and the majority of the team believes in the basic direction.

The same thing holds for schedules. The team has to buy-off on the proposed schedule for a project. Without that buyoff, no one will work to meet the schedules. And then productivity drops.

At the same time, the schedules are aggressive. In the case of programmers this is essential since you can work on a program forever and keep making it better. However, the same holds for almost any job. Figure out the fastest way it can be reasonably done and shoot for that.

VOLUNTARY PERFORMANCE

Finally, there are three key components that make employees at Microsoft want to perform at peak levels.

First, the work itself is, by and large, interesting and exciting. People do a better job if they enjoy what they're doing.

Second, employees want to work there because their work is measured by performance and success. This is equivalent to players who want to join a city league sports team. Their value is measured by their performance and the team's success. And because this is the motivating factor in why employees want to work for Microsoft, they do a good job, which in turn makes them feel better and enjoy their work more. This circular reinforcement is an incredibly powerful tool for boosting both the quality of the product made and the productivity of the group.

The third key component is that peer pressure keeps everyone performing. Your peers expect you to do the best you are capable of, and they are depending on you for the success of the project. In addition, the respect of your peers depends largely on how good a job you do. In an environment so focused on performance and success, that is how your peers measure you. You cannot stand high in their eyes unless you do an excellent job.

In this setting, you have every employee self-motivated to create the best possible product in the quickest amount of time. Nothing imposed by the corporation can match this for productivity.

THE DARK SIDE

There is a downside to all of this. Employees are very wrapped up in their jobs, the success of their group, and the success of the company. The importance of this success lies not just in terms of salary and stock, but in terms of core emotional happiness.

So what happens when they see the company doing something stupid? They let management know—and managers must respond positively. They must not ignore or devalue employee concerns. Not only are the positive effects of the system lost, but productivity drops because of the employees' emotional attachment to the company's success.

Employee concerns must be directly addressed. When there is a major change of direction, Microsoft holds meeting after meeting to present the logic behind the change and to answer questions. (I've thought at times that Microsoft intentionally holds so many meetings because it wants to get people bored with the issue so they'll stop objecting.) Communicating the reasons for change often addresses many employee concerns that were already discussed by management before making the decision. However, the employees don't know that unless they themselves hear the answer and/or reasoning. This also demonstrates that management is thinking, thereby increasing the credibility of the management team. And sometimes an employee will ask a question that brings up issues that weren't considered—possibly something critical to the success of the company.

Never, never try to pull one over on the employees. Employees are smart and know when they are being fed a line. At many companies they don't say anything because that's how the game is played. At Microsoft, however, all they do is point out when policy is a smokescreen and ask for a real answer. Either way, this amounts to a credibility loss for management. (Of course, even at Microsoft, human resources generates meaningless policy sometimes. That seems to be a fundamental law of the universe.

Finally, you are never going to have a change that everyone supports, so you will always have some employees who are upset. That comes with everyone caring about what the company is doing, and accepting this conflict is part of the package.

IMAGINE

There is an easy rule of thumb to determine if a company is effectively managed. Pick the CEO of any well-run company. Got one? If not, think of someone like Sam Walton or Michael Eisner.

Now pick the CEO of any badly run company. Got one? If not, think of the CEO of any major phone or car company.

Here's the question. Imagine what the CEO of the well-run company you picked would do if a nonmanagerial employee asked him a critical question about the company's direction. Almost all such CEOs would answer the question and possibly have questions of their own in return.

Now imagine what the CEO of the badly run company would do. Would such CEOs ever even be in a situation where a nonmanagerial employee could talk to them? How would they answer? Is the corporate culture such that the employee would even dare ask?

It's probably not a perfect measure, but it's a good bet that there is an extremely high correlation between the access and attention an employee can obtain from upper management and the success of the company. A company can't be highly productive and innovative unless the employees buy into the corporate direction and philosophy. And that requires that management sell the employees on that direction and be responsive to their concerns.

7

"SHRIMP vs. WEENIES"

Mike Murray (Microsoft's vice president of human resources) sent out an infamous internal memo that was quickly labeled the "Shrimp vs. Weenies Memo." The crux of the memo, which was approved by Bill, made two points.

The first was that Microsoft was a company that bought inexpensive weenies for food, not expensive shrimp. This was an example, of course, the real message being: Spend money frugally.

The company has consistently followed that penny-pinching philosophy. Bill Gates and Steve Ballmer are probably the only billionaires in the world who fly coach. They fly coach because then everyone else in the company must fly coach too. (As this book is going to press, Microsoft just purchased its first corporate jet—for scheduling reasons.)

There are only two office sizes at Microsoft: regular and double, where a connecting wall is removed. Double offices are not just for vice presidents; they are for senior developers, too. There are no special parking areas, executive cafeterias, etc.

In correspondence with the multimedia division of a major Hollywood film company (which division has subsequently been shut down), I received a letter from the division's contract administrator.

She had her own stationery (i.e., company stationery with her name printed on it). At Microsoft Bill may have stationery that has his name imprinted, but absolutely no one else does.

Since there is no real business need to have anyone within a company have stationery printed with his or her name on it, it's a pretty clear indication that a company has gone the shrimp route when even the contract administrators have personalized stationery.

The second major point, expressly stated, was that when a job at Microsoft absolutely, positively needed five people to complete it, four would be assigned. The philosophy behind this limitation was that only by choking resources back to the barely survivable would work be limited to only those things that absolutely had to occur. And because it is so difficult to find qualified people (see Chapter 2 on hiring only the top 5 percent), this allows Microsoft to accomplish more with the people they have.

Now, neither of these points went down well with most employees. Virtually all of them could point out many examples of other groups that spent too much money or had too many people, while insisting that their own group was exceptionally frugal and needed more resources.

But from a corporate viewpoint, this enshrined an essential philosophy that has allowed Microsoft to continue to stay competitive: Even though Microsoft has billions of dollars in the bank and essentially no debt, money is kept tight. If it were not, that money would go toward things that would make the company slow and inefficient.

And once all the luxuries like ornate offices and personalized stationery are brought in, they continue to be an economic drain with no return on investment. Instead, Microsoft remains in start-up mode wherein money is tight and can only be spent on items essential to the success of the project.

It's a sad fact that at most companies, a significant percentage of the employees do not really contribute to the company's bottom line. However, both to justify their jobs and to have something to do, people will create work. This work generally consists of receiving status, posting status, and having meetings to determine status.

This consumes the time of those trying to get the actual work done, because someone has to originate this status, attend meetings, respond to questions, and continually justify actions. The whole process gets more and more top-heavy until it collapses and the company either goes out of business or, out of desperation, removes all or a portion of the dead weight. And if there is not a continual, conscious effort to keep this bureaucracy to a minimum, it tends to start building up again.

But if there are never enough people available, there is no one sitting around to create a make-work job. The giant savings here is not the direct salary cost of people doing nothing. The savings is the lack of bureaucracy imposed by people trying to create work for themselves.

In short, Microsoft is still a start-up. In a start-up money is tight and is spent only on essentials. In a start-up there are never enough people to do even the essential jobs and never anyone available to do a nonessential job. And in this environment everyone is focused on getting the job done instead of explaining why it can't get done.

Microsoft is simply a start-up that has 25,000 employees, complete domination of numerous market segments, and very deep pockets. This is why Microsoft's growth continues to increase at a rate that is usually only seen in start-ups—that's exactly how it operates.

For those who have wondered why Microsoft's software has always been a bit less than expected, here's a large part of the reason:

There are only enough people to implement the features that absolutely must be included, and there is only enough time to implement each feature in the fastest acceptable manner.

And there are absolutely only enough people and time to test the product to the point where the market will marginally accept it; no more. What is accepted by the market forced Microsoft to increase its testing several years ago but Internet updates are now allowing it to reduce testing once again.

Nowhere is the shrimp vs. weenies approach more visible than when it comes to secretaries. There are no secretaries at Microsoft. Bill and the vice presidents each have an administrative assistant, whose responsibilities are clearly different than those of the typical corporate secretary. These individuals assist in keeping the groups running and are empowered to make a substantial number of decisions on their own.

All the people at Microsoft, Bill included, read their own e-mail and do their own typing. If the CEO of the company does his own typing and reads his own e-mail, then no other employees are going to have a secretary to perform these tasks.

While this reduces costs, more importantly it dramatically improves productivity. E-mail and phone calls go directly between the two people who need to discuss each issue. This eliminates the artificial layers that spring up at most companies that add nothing but slow down both communication and decision making.

As discussed in an earlier chapter, Microsoft does not allow anyone to coast on previous work. In fact, if employees burn out doing a job, however awesome, they are likely to be cast aside once the project is completed because they are no longer of use to the company.

This book does not claim that absolutely everything Microsoft does is proper, or even, for that matter, effective. This approach—using up employees and casting them out, albeit with their stock options, when they are no longer needed—is clearly ruthless. When taken to the extreme, as it is at Microsoft, it is a heartless treatment of workers. However, it is also an approach that is more profitable than carrying dead wood. And the opposite extreme, carrying employees forever because they once did something useful, is a damaging luxury that few companies can afford any longer.

Bottom line: If an employee is no longer needed, then regardless of what the employee has accomplished or what the employee is capable of, he or she is cast aside. If each program is run on weenies and not shrimp, then this is essential. People, no matter how good, are not kept if not needed. If this isn't done, then programs slowly staff up with excess people because there is no incentive to get rid of extra resources.

Once again, the end result is the equivalent of a small, entrepreneurial start-up that never has enough people to go around—and, at the extreme, has to turn away even qualified people because it can only afford one less person than is actually needed. At the same time, with rare exceptions, there are other groups within Microsoft that have positions to be filled. A position may not be the exact job someone wants, but there are opportunities, so things are generally not as heartless as they seem. Further, people are forced to make the decision to either learn a new skill that matches the company's existing needs or leave. The alternative is keeping people with obsolete skills, and once again, that's not just an extra expense but a drag on the company's productivity.

8

SIZE DOES MATTER

E veryone sees Microsoft as this big monolith, working like the Borg as a single entity, swallowing
everything in its path. This impression couldn't be further from the truth. Microsoft is not a single, large
company; rather it is a collection of small, independent
companies. The primary job functions at Microsoft are
creating, testing, marketing and selling software.

And, amazingly enough, these functions are
largely performed separately for each and every project. The greatest cross-fertilization occurs at lunchtime when people in disparate groups that know one
another eat lunch together and talk about what they
are working on.

Why is this? Because Microsoft realizes that large

companies become stagnant. The process becomes more important than the work, and as this happens, productivity slows to a standstill.

TO EACH ITS OWN

First off, the same method is not the best for all projects. A single bureaucratic structure, no matter how well designed, will never work as well as a structure designed specifically for each project. By remaining a set of distinct groups, each can design the process that works best for it.

And at the same time, because the group has designed its own process, it can amend it any time it wishes. A process imposed from above has to be appealed and even if the appeal is approved, that approval takes time and energy—time and energy better spent on the project itself.

And because it is a process established by the group, it is owned by that group. And a process owned by the group is much more likely to be followed. This is understood even by those kindergarten teachers who ask their students to determine the rules. Does this mean that many corporations assume their employees are less responsible or intelligent than the average kindergartener?

SOME ELEMENTS ARE MONOLITHIC

Some elements do need to be monolithic, even at Microsoft. All of the services such as mail, phones, and MIS are run by a central body. However, every one of these services could be handled by an outside contractor instead of internal employees and Microsoft would be the same company it is now.

In addition, where the company meets the public you will find a single organization. The public does not want to have to figure out whom to contact within Microsoft, and so divisions such as public relations, product support (technical support), and customer service are global and one ad agency is used to design advertisements for major publications to ensure a consistent style. The purpose is to allow Microsoft to present a unified picture to the world.

But the clear priority is to keep things as broken up as possible. Development teams function independently, as previously discussed. Marketing plans are created within each project group. (Marketing materials must be approved by a central group for legal purposes, and, as just mentioned above, for the purpose of assuring some level of consistency, but the ownership of the materials is held by the project.)

The governing concept is that, whenever possible, the work be kept at the project level. By keeping each project functioning as an autonomous group, strong ownership is also retained.

SHARING IDEAS

Virtually all job creation, technical innovation, and exceptional productivity in the U.S. economy comes from or is found in small companies. Be it the independent American spirit or the bureaucracy generated by large companies, small companies are where you find productivity.

This helps explain why Microsoft continues to grow at such an incredible rate. Supposedly only small companies can grow at this rate, and everyone is waiting for Microsoft's growth to match that of a large company. But if Microsoft continues to operate as a small company internally, then it can continue to grow at the rate of a small company. And that means it can continue its present growth rate unabated.

If Microsoft were just a collection of small companies with Microsoft's financial reserves and marketing muscle behind it, that alone would assure continued success. But there's more to it.

Bill and the other senior executives have a very detailed view of exactly what is happening throughout the company. No project is created or killed without Bill's active participation in the decision. Therefore, while each group controls its destiny, it is also part of a grand, strategic plan.

This is not to say that every project must fit the plan. Many projects are started because they might lead somewhere interesting. However the farther they stray from the strategic goals of the company, the less likely they are to be approved. And the more critical they are, the more attention Bill and the other senior executives give to them.

Employees at Microsoft also pay attention to what other groups are doing. When they see a good idea elsewhere, they grab it for their group's use. In addition, employees at Microsoft are not shy about trumpeting their success. They will pass on what has worked well for them to others via informal e-mail to friends, white papers, or presentations. In some cases the management from one group will be asked to present their new process to the management of several other groups. Through the presentation and Q&A following, these other groups can learn what the new process is and why it works.

When there is a significant problem affecting numerous groups, Microsoft will bring together representatives from various groups to figure out how to solve it. This group is generally composed of people who are highly respected within the company. They will propose solutions to the problem, and largely because of their reputations, virtually every manager in the company will seriously consider their recommendations. By this means alone, many new concepts are voluntarily picked up by different groups.

For example, when Microsoft Office first became successful, all Microsoft applications had a different interface. Each project team realized that it needed to standardize on a common interface, and so a group was established to work together to hammer out a common interface. The only real outside influence was senior management's insistence that the groups reach a consensus.

KEEP IT SMALL

The real point is not that smaller is better. The point is that smaller is essential. This is critically important. If a company is not organized as a loosely knit group of small companies, it will fall prey to a company that is. Or even to a small company a fraction of its size. (Look at Amazon.com's valuation versus that of Barnes & Noble.)

And many who do try to keep the company a group of independent entities don't break it up enough. There are many companies that have separate business units but retain numerous "essential" required processes. The result is that the units are not a group of independent companies but only slightly autonomous parts of one big company.

The delicate balance between independence and control must be maintained. Companies cannot break apart and lose control of the pieces. The result would be a number of independent fiefdoms all working at cross-purposes. What is required to achieve this is a very strong CEO and a cohesive senior management team. They must be experts in the company's business and have strong personalities that can guide the independent groups in accordance with the company's strategic plan.

9

BILL IS WATCHING

I once asked someone why he left Microsoft. He said it was because every day it was clear that it's Bill's company and everything was going to be done Bill's way. His point that it is Bill's company is dead on. Bill's approach, his philosophy, and his strategic vision permeates the entire company. And as the individual who left demonstrates, you either get with the program or leave.

This does not mean that everyone is a robot marching along as a mini-Bill clone (although Bill would probably prefer that). But it does mean that everyone accepts most of the precepts and embraces a reasonable number of them. The result is that Bill controls Microsoft to a much greater degree than does the CEO of any other large company. (The exception might have been Walt Disney.)

Every month the lead on each project e-mails a status report to Bill and copies everyone in the food chain between Bill and the lead. The report is not sent by a vice president or a department head. It is sent by the person managing that project.

This person has the clearest picture of the state of the project and understands what problems are surfacing. More importantly, because the person is someone who is not in upper management, he or she is much more willing to state what is wrong rather than only reporting good news.

Lyndon Johnson, when asked how much he worried about his enemies, replied that it was his friends that had him up all night with worry. As president, none of his friends wanted to tell him unpleasant truths. And so problems would get much worse before he would become aware of them.

These monthly reports have a set format, and that format is designed to clearly report two major items: the current status of the project and the major problems it is facing. It tells Bill what is happening and what to worry about—in detail.

Virtually every software company includes in its status report on each project something generally called "the top 10 risks." There are not always 10 items on the list, but the purpose is clear—determine what the major risks are to a project and how to address those risks as early as possible.

The important thing here is that in most companies, upper management has absolutely no idea what is going on. In addition, it doesn't want to know what is going on. After years of being told everything is OK, finding out how screwed up things truly are is way too scary.

So decisions made by upper management, at best, will not hurt what is actually being done at the project level. However, many times decisions at the upper management level will actually exacerbate problems because the managers at this level aren't even aware of these problems. Furthermore, because the top priority at the average company is to keep problems hidden from upper management, resolving problems gets less attention; and at times, resolutions are not even attempted because that would bring the problems to the attention of upper management. Now compare this with what happens at Microsoft. First off, all the managers in the management chain, up to and including Bill, are aware of every major problem in every project they are responsible for. So the decisions they make take into account *reality*—the various problems they are facing.

Second, there is pressure from above to address problems and support from above to help solve those problems. This encourages people to fix the problems they face quickly, and it provides assistance where needed to reduce risks.

BILL IS CALLING

Almost every Saturday morning each vice president is sitting in his or her office. The sun may be out (rare for Seattle), or the kids may have a soccer game. There may not be any critical issues to resolve. But the vice president is there. Why? Because every Saturday morning Bill calls each and every one of them and spends half an hour discussing the various issues their department has. Every Saturday they dive into the details of the department one-on-one.

And Bill does dive in on the details. This is not a glossy pass over the department. While half an hour does not seem like a lot, it's time spent consistently every week. And since Bill is already familiar with the details, the time is spent productively discussing changes and problems.

The result is that Bill keeps control of each division and everyone is kept working toward the same goals. Yes, it means that vice presidents at Microsoft in some respects have significantly less autonomy than at other companies, but that lack of full control is better for the company.

BILL COMMUNICATES THE MESSAGE

Every year the entire company is bused to a large building in downtown Seattle (each year it's a bigger building because the company keeps growing). This is the annual company meeting. It's part religious revival and part business report. There is always a comedian or some other entertainment to get the group laughing and cheering. The senior executives follow, explaining the present state of the business and the strategic direction for the future. It's detailed and direct.

The meeting is open to employees only. Everyone knows that the topics, after all the skits and kidding, are serious business. These are the marching orders, and all the employees take them into account in every part of their job.

Afterward booths are set up showing the interesting new products in development. In some cases you can see things that won't be shown to the rest of the world for years. It is a significant effort, and the company invests a lot of time and money in these meetings. They are not just rah-rah, feel-good, propaganda opportunities. Instead they are a focused effort to communicate the company's goals—presented in a very entertaining fashion.

The result is that everyone understands the strategic direction of the company and the major tactical efforts going on to achieve that strategic goal.

And there's more than a single, annual meeting. Each division has at least one additional meeting a year to discuss strategy and products for that division. The same goes for each business unit, department, etc. And in each case the information clearly states where that group is going (and usually in an entertaining manner).

What is the outcome of all this? First of all, no one goes to sleep because the meetings are kept entertaining. This is clearly an important requirement for every major event.

Second, the meetings treat the attendees as highly educated and motivated employees (they are) who will make good use of knowing where the company is headed.

The result is that all the employees know what Microsoft's strategic goals are and they take that into account when doing their jobs. So you end up with 25,000 people who are all trying to accomplish the same thing—Bill's strategic vision.

BILL WANTS TO KNOW

And then there's the "billg" meeting. This is an official review with Bill. The focus might be on an entire project, of which there are many over the lifetime of a larger project, or it might be on a specific issue.

These meetings are not your typical presentation to upper management where everything is quiet and refined. The meetings are direct, focused, and downright rude at times. One of Bill's favorite replies when he thinks something is not well thought out is, "That's the dumbest thing I've ever heard."

But this is not an event for Bill to whip a defenseless subordinate. The others in the meeting are expected to fight for what they believe should be done. As one vice president said, "Bill respects no. You just have to be able to back it up."

The meetings are loud and contentious in many

cases, but they are by no means one-sided for those prepared. However, they are truly awful for the unprepared.

Early on when Dave Cutler (the newly hired architect of Windows/NT) had his first technical review with Bill, he did not take any of his developers with him. After all, at any other company you would run all questions from the CEO through the manager of the project.

Bill started asking detailed questions about specific parts of the code. Dave told him that he didn't know but would get back to him. Bill ended the meeting prematurely and told him not to come back unless he brought the people with him who could answer the questions.

This illustrates two very critical points. First, it wasn't that the people doing the actual work *could* talk to Bill. They were *required* to be there. He wasn't going to do the review without the people present who could speak directly on the work being done. In practice, anyone doing a product review at Microsoft, right up to the CEO, talks directly to the person doing the work.

Second, Bill, who is the CEO of the second most valuable company in America, talks directly to the people doing the work when reviewing it. (Sam Walton, who spent a lot of his time walking in his various stores, McDonald's founder Ray Kroc, and Walt Disney did similar things. All these people ran very successful companies.) But how many other senior executives do this?

It's also more than just talking directly to the employees doing the work. It's understanding the work being done and understanding it in detail—and retaining that information from one meeting to the next.

It is clear that Bill remembers arcane technical details from previous meetings on the project under review, as well as from other projects. These are not meetings where information is being passed on to a senior executive who cannot follow it. This is the presentation of information to someone as astute, both technically and businesswise, as most everyone on the team.

At one meeting on a future O/S project, those attending listened as Bill and Mark Zbikowski spent 15 minutes arguing about the best approach for the base object-oriented file system. This was a detailed technical discussion, and it was clear from Bill's arguments that he understood the system and how it was designed.

Equally important is that it was a discussion between Bill and Mark. It was not Bill looking at it, understanding it, and then issuing a directive. It was a major debate. Bill can often be convinced to change his mind or at least defer to the other person's judgment. But because of Bill's level of knowledge, the person discussing the issue with him better have very strong arguments.

The outcome is that major decisions are made quickly and intelligently by someone who understands what is going on elsewhere within the company. And because these decisions are made by Bill himself, after the meeting the project can move ahead instead of waiting to see if the decision will change as it is presented to others. Further, because Bill understands what is happening throughout the company, the decisions are generally the correct ones for the strategic direction of the company. (This same type of review occurs at every level in the management chain.)

Contrast this with CEOs of even average mid-sized companies. They don't have the background to make decisions of this kind, and they don't have the information about what is happening elsewhere to make an educated decision. Decisions end up being made by various groups, each with its own direction—not a good situation at all.

BILL IS SLUMMING

In the early days it was not uncommon to return to your office and find that Bill had not only been reviewing your code, but had improved it. He still cruises the halls of a given project once in awhile and will pop in at random to see what someone is working on. This does not deliver a big-picture view of a project. But by talking with 3 or 5 or 10 of the individuals doing the work on a project, he receives a very detailed view of pieces of that project. Further, he gets an excellent overview of the general state of the project and of the people on the project.

More to the point, Bill is getting this information directly and one-on-one. Try to imagine the CEO of any other major corporation doing this. At most companies, they rarely leave the executive floor and are never found where work is actually being accomplished.

Without this ongoing effort to see what is happening at the individual level, the CEO loses the feel of the company and loses the close control found at Microsoft. Furthermore the fact that Bill could show up in your office at any time is a constant reminder to everyone of how closely he is watching and directing the company.

During World War II only one American general was killed in action (General Rose, who was commanding a tank in an assault in Germany). The reason for this was that most of the generals stayed way behind the front line. By staying behind they were safe. But it also meant they had no feel for the battle and it led to numerous disasters, the biggest probably being the needless battle for the Hurtegen Forest.

Besides Rose, the exceptions included Generals Gavin and Patton, who did put themselves in harm's way to stay in touch with the battle. (Patton did this to the extent that he was reprimanded by Eisenhower for needlessly endangering himself.)

Yet by and large, these were the best tactical generals America had. They consistently beat superior numbers of Germans while on the offensive.

MANAGEMENT IS WATCHING

And none of this is a Bill-only situation. Other executives within Microsoft do the same thing, although they all have their own style. Presentations to various executives about what is happening, along with immediate decision making, are common.

Managers are expected to, and do, have a good picture of what is happening below them, and they drive the major decisions for their group or push them up to the appropriate person above when necessary. To refer to the Saturday morning phone calls Bill makes, since both Bill and the vice president have this detailed understanding of the vice president's division or business unit, they can quickly focus on the major issues at hand.

BILL IS LISTENING

And finally, throughout the company, on various projects and at various levels, are the people who knew Bill back when the company was much smaller. These people can shoot Bill an e-mail and he will pay attention to it because he knows the sender and can measure the person's credibility.

Bill acts on these e-mails as he feels is appropriate. He may ignore some, but other times he will agree that there is a situation he needs to look into and he will do so.

Software is somewhat unique in that very senior people will choose to remain developers or go from management back to development. And because Bill started Microsoft with two people, everyone from the early days knew Bill personally.

This all means there are many individuals who are simply senior developers on a project who can e-mail Bill directly and get a response. And this means Bill (and the other senior executives who all have their own web of people they know within the organization) gets constant feedback from throughout the company.

MANAGEMENT IS IN CONTROL

The individual steps to take control are not difficult. The result is that not only do the senior executives know what the company is doing and what the problems are, but they are also actually controlling the company.

The alternative, found at all too many companies, is that senior management has a poor view of its own company and can only exert some influence on it.

Guess which company will be more successful?

10

ESPRIT DE CORPS

Almost every manager understands that morale affects productivity. Many managers try to work around the issue by taking the easy route: hiring pretty much anyone, paying him or her the least amount possible, and keeping job responsibilities to an absolute minimum. All they succeed in doing is bringing productivity down to the lowest common denominator.

Other managers try to improve morale and thereby increase productivity. Some do this better than others but are often limited by their own skills or by restrictions imposed by the company.

A few companies actively try to improve morale (most only pay lip service to it), and these companies have varying degrees of success. The sad fact is that most companies don't even try.

But this chapter is not about morale. All companies have morale. Some have low morale. Some have high morale. But all companies have a level of morale.

This chapter is about esprit de corps. True esprit de corps is generally found only in small groups of highly trained and highly motivated individuals. It is found in sports teams that work well together and in the best military units, although generally only up to the platoon or company level.

It is also common at most start-ups. Start-ups are composed of small groups of people all focused on a common goal. Esprit de corps can almost be viewed as an essential component of a start-up, because of the advantage it confers on it. And if a start-up cannot build esprit de corps, then how can it succeed?

AT MICROSOFT

At Microsoft you have a very strong sense of esprit de corps. And as a result, morale is incredibly high—another tremendous competitive advantage for Microsoft. How is a company going to compete against a company that has 25,000 people who are all charged up to accomplish anything and all focused on achieving the same goal?

The larger the company, the harder it is to keep this spirit. A lot of it comes from the items discussed in previous chapters. An ineffectively managed company is never going to have great morale because you can't have great morale if you are managed by losers.

At the same time, effective management and success in the market are not enough. It takes a lot more to build up the level of esprit de corps found at Microsoft. These extras are what are covered here.

OWNERSHIP

People do not knock themselves out for things that they do not have ownership or a personal stake in. However, for something that is their baby, their creation, people will do almost anything.

All the employees at Microsoft own their piece of the project. If they are programmers they design the piece, code it, and test it. Someone else does not do the design and then pass it on for them to code. The programmers are responsible for the entire piece.

The same holds for program managers, marketing directors, etc. In each case it is the responsibility of the individual employee to determine what needs to be done and then do it. Obviously, all these people have to work together, and the constraints of other pieces will sometimes limit choices, but each person still owns his or her piece of the puzzle.

FREEDOM

With freedom comes responsibility. The employees at Microsoft have the freedom to determine how best to accomplish their job. By giving them the responsibility of making that decision, the employees strive to make the best possible decisions rather than waiting for others to do it for them.

This does not mean employees work in a vacuum. Key decisions need to be run by managers. The major design is presented to coworkers for review. On the marketing side there are committees whose approval must be obtained to run an ad. But even with this approval process, it is the job of the employee to come up with the approach and to obtain the approval. And if an approach is not approved, then the employee must come up with another approach. Responsibility is never lifted from the employee.

This responsibility in turn confers upon the employees the freedom to make their own decisions. Employees are free to select whatever approach they think is best as long as their coworkers cannot find any major problems with it.

The review process is generally an iterative process where the design will be presented and others respond with suggestions. Based on the input received, the individual will improve the design. And since the goal of everyone is to create the best possible design, it is generally a cooperative process.

FOCUS

Equally critical to esprit de corps is focus. Just as no one has ever died giving his or her life for *two* causes, no employee can give fair and equal attention to two different projects. With rare exception, each employee at Microsoft has one major task. For a vice president that task may be to take over the entire desktop application market. But it's one task.

This focus gives each employee something to live and die for. Employees can live, eat, sleep, and breathe their project. And they do. This is not hyperbole. People at Microsoft will, at times, do little else other than sleep and work. And they often work these hours of their own volition, not because they are required to.

This focus also allows employees to be more productive. An employee who shifts between four projects during a single day generally spends over half of that time just switching gears. (The more technical the job, the more time spent switching.) Creating a highly productive environment once again helps create esprit de corps because people are happier when they can be productive.

A TEAM IS A TEAM

At Microsoft teams are created for each project, and people are assigned to the team for as long as they are needed. Over the course of their assignment, they may not be needed full time but they are kept on the team anyway. And they stay there for the course of the project.

While this violates the shrimp vs. weenies concept, it is a critical factor. If people know they are moving on soon, they will not feel a part of the team, nor will they be dedicated to the outcome of the project. Employees at Microsoft are not on one team one week and on another the next. They know they are there for the duration.

When every team member is there for the course of the project, then everyone comes to own that project. In fact, if many of the team members were transitory, then esprit de corps would be impossible.

Because all the members are in it together, and they are dependent upon one another for a success, then they can build up the necessary spirit.

MEETINGS AND RETREATS

When I first started at Microsoft, it seemed like every other day was spent at either a division or higher level meeting or retreat. Work needed to be done, and we were spending all this time being told where we were going and why! It soon became clear, however, why this time was invested.

If done correctly, the information disseminated not only allows everyone to make decisions based on where the project/division/company is going, but also brings everyone together so all employees become part of the overall corporate team.

A key point: These meetings are entertaining, not just with marginally funny skits thinly veiling a corporate message. They are pure entertainment, many times skewering upper management or management goofs.

As mentioned earlier, entertainment gets everyone through the meeting without falling asleep. But it also makes it an event for the group. Everyone is laughing together. Everyone has something to talk about the next day. Participating in a common event is a critical piece for building esprit de corps within an organization.

Most of these meetings end with a party with free food—good food. This gives people an additional chance to talk together and build up the team spirit.

One year for the systems (everyone working on operating systems, etc.) retreat one of the standard-issue training companies was hired to do a bunch of team-building events. It was a dismal failure. Some of the employees flat-out refused to participate in the events. Others argued endlessly with the team leads from the hired company. It was not repeated.

The meetings have to be very carefully planned. True team building is good. But the techniques peddled by many professional team-building consultants not only are ineffective, but will be rejected by an independent group of highly motivated people.

PRACTICAL JOKES

At a seminar I attended recently, a management consultant advised that humor is inappropriate in the workplace. Baloney! It is absolutely essential. A cold, sterile, humorless workplace can never develop esprit de corps.

> *One day a team member was leaving on a week's vacation the same day that the rest of the team was getting the latest version of Visual C++. (This was back when VC++ came in a large box, with multiple boxes inside.)*
>
> *I e-mailed all the members in the group and told them to put their boxes (separated, of course) in his office. When he returned, his office was filled almost to the ceiling with empty boxes. He was pleased to no end!*

Running around shooting Nerf balls at each other in the halls may not seem like a productive use of time. However, something very interesting is happening during these Nerf ball wars. People at work are having fun. Not a little fun, but rather an absolute blast.

The office becomes a place where you have fun. And you have it by doing things you can't do elsewhere. (Where else can you find 30 people to have a Nerf ball war on the spur of the moment?) This is what cements the group. And it can't be done without fun and even practical jokes. This is absolutely essential.

It's more than the jokes themselves. The fact that this can be done at work is itself critically important. It tells the employees that the company is supportive of them enjoying themselves in the workplace. And that as long as the job is getting done, that's all that matters.

Ask yourself: If you absolutely knew it would double productivity, would your company allow Nerf ball wars in the halls? And if the CEO of the company got pegged, would it be OK? The answer for most companies to this is a *big* no, and it's that answer, not the actual Nerf ball wars (which got annoying after awhile), that makes the big difference.

Whatever the specific activity—be it fighting Nerf ball wars, or Supersoaker wars, filling offices with boxes or balls, it doesn't matter. What matters is that when groups do these activities, they are, on their own, performing their own team-building exercises.

And because these are things they decide to do on their own, things they generally can't do elsewhere, work becomes a place where they have fun—as a team.

AWARDS

There are not a lot of awards given out at Microsoft. This is the opposite approach to that taken by a lot of companies that try to distribute awards and special bonuses for various small things.

At yearly division level meetings, there are generally awards for a few people who did an outstanding job. The award in these cases is additional Microsoft stock. Occasionally there are extra grants of stock for completing a really impossible task or for signing up to do something that is a personal hardship. But again, these are rare. The one award that is given, and does matter a lot, is the award a project gives each member when the project is shipped. These matter because they attest to the fact that you have shipped a finished product.

So what's the point? The point is that awards are definitely not necessary to have a successful and productive company. In addition, it may be that awards are indicative of an organization that is incapable of creating esprit de corps. Awards are a cheap way of providing incentive without imperiling any of the existing practices of a company.

While awards have been proved to be an effective motivator in many cases, the question every company needs to ask is: What are we doing to create esprit de corps, aside from giving out awards? (As opposed to pointing to the awards and deciding that no further effort is necessary.)

STOCK OPTIONS

Stock options were purposely left to the end of this chapter. They are not the key item to building esprit de corps. Yes, if Microsoft eliminated them, many people would leave—because they do want to be paid well.

But that's it. Employees want to be paid well. But the other elements, an enjoyable place to work with very smart people and the opportunity to be part of a cohesive team, count for a lot more. Unquestionably, employees need to be well compensated. And they need to be tied in to the success of the company. Stock options are easily the most effective way to accomplish this.

Stock options are equally important in keeping

people motivated when they are upset with the company. When work becomes very frustrating, waiting for the next vesting keeps an employee going.

But realize that while stock options are a necessity nowadays, they are not sufficient by themselves as a means of building up morale. And they are, in fact, a very minor component of boosting esprit de corps.

On the flip side, not delivering stock options is an excellent way to ensure morale will remain low. As Sam Walton demonstrated, even for retail store clerks, stock options can dramatically boost morale and thereby productivity.

CHARGE THE RAMPARTS

The tough thing about esprit de corps is it's so hard to get your hands on it. A company could implement everything listed above and still not have it. Another company could take a totally different approach and succeed.

Each company needs to find its own approach. A downtown law firm cannot have squirt gun fights in the halls because its clients would never understand. But gaining that elusive feeling is worth almost any effort, because the boost in productivity and focus is phenomenal. The U.S. Army consistently beat larger numbers of better armed German troops in World War II. A very large part of this was due to the superior esprit de corps of the American troops.

And if your company does not have this elusive spirit while your competitor does, then you cannot win. You may hang on for a time. You may use connections and existing market share to delay the inevitable. But eventually you will lose.

With esprit de corps, you can take market share away from your competitors and they won't even understand why. And if they don't get it, you will be able to continue taking market share until they leave that market or go out of business.

11

STOP THE INSANITY

BADGES? WE DON'T NEED NO STINKING BADGES!

An unnamed high-tech company moved one of its local divisions into a new, larger building. Security had been pretty much a nonissue up until then. No one had badges, and there were no card keys. Step 1 was to get everyone badges just as the other locations had. It was new, but no big deal.

Then one morning people came to work and found in their mailboxes a four-page memo on proper badge procedure. Now, first of all, the level of detail describing how and where to wear the badge was totally ridiculous. People were being treated like complete morons, and, not surprisingly, people were ticked off.

To make things even worse, this directive was not handed down through management so that a better spin could be put on it. Instead it was given directly to each and every employee so managers had to scramble to try to minimize the damage.

The memo was never shown to senior management prior to distribution, so the people at the top didn't know what was happening. And because nothing was heard from senior managers, employees assumed they had approved the contents of the memo and its distribution.

Clearly, the company did something really stupid and for no good reason. (I say no good reason because no one is going to remember four pages of rules about wearing badges.) Does Microsoft manage to avoid this type of inane garbage? By and large yes. To speak specifically of badges, with a lot more employees than this other company, Microsoft takes the employee's picture and hands the employee his or her badge. That's about it. It's assumed an employee can figure out how to use and wear a badge. [Microsoft does state that security guards can stop you and ask to see your badge if it's not visible. So for those who keep it in a shirt pocket (easier to pull out and swipe), once or twice a year they will be asked to show their badge—but that's it.]

Occasionally something stupid is done at Microsoft, but generally whatever it is gets fixed or addressed quickly. One day notes appeared taped to all of the office doors, along with a campaign-type button with Bill's nose and mouth and some standard corporate line on it. Someone (a regular-type employee) e-mailed Bill asking why money was being wasted on this stupidity. Bill replied that it shouldn't be. Two hours later the notes and buttons were removed.

In other words, the company fixed it. Occasional nonsense becomes tolerable if you see the company undo the deed when it becomes aware that it's stupid.

An awful lot of the junk that goes on in companies seems to be directed at ensuring that for even the stupidest employee, there will be rules to guarantee that they act exactly how the company wants them to.

Microsoft generally takes the other tack. It assumes that all their employees are smart and relies on the employees to make smart decisions. And if an employee doesn't, the company deals with that employee rather than impose needless rules on all the others.

NOT ANOTHER MEETING!

Meetings—the biggest productivity draining, life sucking, bureaucracy-building force in the universe.

Darth Vader is nothing compared with the paralysis that meetings can bring to a previously productive unit.

But this force can be harnessed for good. Some meetings are necessary. The trick is keeping them under control.

It's rare to attend a meeting at Microsoft that's a waste of time. (Certainly there are some but they're definitely few and far between.) The meetings held there are almost always necessary.

Effective use of e-mail eliminates the need for a lot of meetings that would otherwise occur. E-mail is the lifeblood of Microsoft. Through e-mail most minor and many mid-level issues are decided.

And hallway conversations between the key three or four people involved solve a large number of other issues. It's a lot faster and more efficient to just grab the needed few and decide something rather than scheduling a meeting on the subject.

The meetings that are held stay focused. They do not wander all over the place. They concentrate on the key issues and don't get bogged down in unimportant details or irrelevant discussions.

Most critical of all, meetings are held to make decisions. As noted earlier it is considered unacceptable at Microsoft to leave a meeting with issues undecided.

There are times where there is no way around having to wait for a decision because of incomplete information. But this is done only if absolutely necessary. And if the delay occurred because an individual did not bring all the information he or she should have, that individual will probably never make that mistake again!

SET INFORMATION FREE

One of the signs of an ineffectively run company is one in which information is viewed as power. People keep information to themselves and let it out in little dribbles only when absolutely necessary.

Microsoft does the opposite, disseminating status, schedules, designs, architecture, direction, etc., as widely as possible. And this information is followed up with more information when it is requested.

If an employee at Microsoft did not disseminate information, others would assume that either the information didn't exist or the employee was hiding bad news. In other words, employees who hold onto information are setting themselves up to fail.

Because information is disseminated so openly, there are no artificial barriers placed in the way of people who need the information. This makes the system a lot more efficient and removes the levels of bureaucracy that information protection requires.

SINCERITY

Everyone has met the type: the manager who always has that phony smile or pasted-on look of feigned interest. And in oozing all of that fake sincerity, the underlying message that comes across is that the person is actually insincere about everything.

At many companies this seems to be the predominate management style. It's the style taught by most management classes: sincere insincerity.

Even Microsoft has a number of these management types (it seems there is no getting away from them anywhere), but it has a lot less of them than most companies do.

People at Microsoft are focused on getting the job done. Insincerity gets in the way of that goal. An insincere manager is less likely to be effective, and senior managers at Microsoft demand an effective manager.

Bob Lewis, an *InfoWorld* columnist, wrote that in fat times corporations want the smooth talkers that say everything is OK. But when a company is in dire straits and needs to produce, then it turns to the employees who can actually get the job done.* Microsoft's trick is it turns to the employees who do the best job even though times are good.

EAT YOUR OWN DOG FOOD

At one of the major car companies (and possibly all three), the senior executives are given a new car every two years. They park in a heated garage at work. If there is a problem with the car, an employee takes care of it for them and finds them a loaner if necessary.

These guys have absolutely no concept of the experiences the consumers of their product go through. They do not have to put up with the insanity of a brand-new vehicle going in to be fixed four times in the first two months (which happened to me and our Jeep Cherokee), topped off by the service department's "tell someone who cares" attitude.

At Microsoft all the employees eat their own dog food. When a product goes into beta, (a fully functioning version of a software program that is put into distribution so that initial users can report on any "bugs"), the entire company switches over to using it,

* Bob Lewis, *InfoWorld;* "IS Survival Guide: Workers with Polish Climb the Corporate Ladder—but Others Get the Job Done, etc . . ." March 30, 1998.

including senior management. When the latest version of Word is shipped, Bill and the others in senior management have already been using it for months.

At one company meeting Bill was asked if senior management used programs when they hit beta. He replied that he did and that he believed that he had tested solitaire more than any other employee.

This eliminates a lot of the usual "exaggeration" that surrounds a product's readiness to be shipped. Everyone has been using it. This makes it difficult to use smoke and mirrors to pretend it's in a state it isn't.

And yes, products are shipped with bugs in them. And despite protestations to the contrary, most everyone at Microsoft is familiar with these bugs. But the end result is that the decision to ship is made based on the true state of the product.

In a bureaucratic organization, not only is the product generally shipped with senior managers having no idea what condition it is actually in, but they many times do not ever learn about the condition afterward either. Look at the car companies. The senior managers probably consider their customers a bunch of whiners because they themselves have no problem with their vehicles and have not been inconvenienced. The bureaucracy insulates them from the real effects of using their product so they have no idea what's really going on.

THE NAVY WAY

The U.S. Navy has a guideline that no matter how critical a regulation might be, don't make it a regulation if it's not going to be followed. All you do is lower respect for all other regulations when you have one that is commonly flouted.

It has always seemed logical to keep essential procedures down to a couple of sheets of paper. Because if it's more than that, no one is going to remember them all anyway. Microsoft practices this principle. The rules are few and far between. And the ones that are there by and large make sense. And because the rules are so few, employees follow them because they can remember them.

Or to put it another way, many Dilbert cartoons do not strike a chord at Microsoft.

12

HOME
AWAY FROM HOME

I magine for a moment that your company wants to make its employees as comfortable, productive, and happy as possible at work. How would you design the office environment? Where should you look for inspiration? Look to the home. People design their homes to be a place that is comfortable and enjoyable. They design home offices to be places of enjoyable productivity.

OFFICES, NOT CUBICLES

Microsoft has offices, not cubicles. (Cubicles have not sold well in homes.) Virtually all full-time employees have their own office. Temporaries and summer interns share an office, rather than having their own. However, even there you would be hard-pressed to find more than two people assigned to an office.

People have their own private space. It's theirs. They can close the door, turn up the music, adjust the lights and work. Because employees have their own preferred working environment, they can set their office up to match their specific needs.

Contrast this to cubicleland, or that even worse alternative, an open pit area with no dividers. (Why is it that the people who insist that cubicles or open pits are the most effective always put themselves in an office?)

In a cubicle environment you cannot make too much noise or change the lighting. You cannot disrupt things in any other manner. You clearly do not "own" your work area.

IT'S YOUR OFFICE, NOT THE COMPANY'S

Your office is yours to do with as you wish. You can put heavy metal posters on the walls, lava lamps on your desk, and Dilbert cartoons on the outside of your door if you so choose.

There have been numerous, extensive studies on the differences in the productivity of programmers housed in offices versus those in cubicles. The studies determined that the difference in productivity is on the order of a factor of 2.5.* This means if you have 40 programmers in cubicles and you need 100 programmers total, you can either hire an additional 60 programmers or put the existing 40 programmers into offices. Cubicles are easy to justify financially because a company can point to the direct saving in office space expense. What is not considered, however, is the indirect cost of the horrendous loss in productivity and the resulting need to more than double the staff.

Cubicles have an even greater, indirect cost: Employees hate them because they are dehumanizing and inefficient. They're right on both counts.

* Steve McConnell, *Software Project Survival Guide,* Microsoft Press, Redmond, WA, 1998, p. 45.

While the majority of people just hang a couple of pictures on the walls, the fact that decorating is up to the occupant of the office means that the only reason the occupant hasn't done more is that he or she hasn't taken the time, not because there is a rule against personalizing the office.

What does all this do? It makes the office the employee's, not the company's. In cubicleland, people are always trying to personalize their cube as much as possible, but they can never hide the fact that it's not theirs and they are in an area designed by the company. At Microsoft, the office is yours, not the company's.

IDENTICAL OFFICES FOR ALL

Everyone except senior architects (the top rank for programmers) and senior vice presidents has the same size office. (Architects and VPs get a double office.) This eliminates all the petty jockeying that occurs when someone gets promoted. If a double office is good enough for Steve Ballmer (now president and a multibillionaire), then it's good enough for everyone else.

The same goes for furniture. All employees, regardless of position, select office furnishings from

A classic is the true story of the guy who worked for a large insurance company. You were allowed a single picture of your family on your desk—no other personal touches.

As a joke he bought a frame that had a picture of a family in it and replaced the photo on a coworker's desk with that picture. He left a note explaining that the coworker's family photograph did not meet the corporate standards so the company was replacing it with an acceptable picture. *

Here's the sad part: Everyone believed it. A depressing commentary on morale at that company.

* Tom DeMarco and Timothy Lister, *Peopleware: Productive Projects and Teams,* Dorset House Publishing Co., New York, 1987; pp. 38–39.

the same catalog. Microsoft has just one type of desk, table, bookshelf, and chair. Desks come with or without returns, the tables in various lengths, and the bookshelves in various heights. But they are all the same series from the manufacturer—for everyone at every level. Employees choose how to furnish their own office.

Some select just a desk and chair; some choose two or three tables and no desk. Instead of the company forcing people to accept the company's idea of the optimal set of furniture, people get the furniture and setup they find most effective.

Once again, this allows employees to work more effectively because they have the physical environment that works best for them and it enforces the concept that the office is the employee's because they determine what furniture will be in it. Further, because you don't get to pick from a nicer line of furniture if you advance a level, it keeps people focused on work instead of perks.

WINDOWS

As many offices as possible at Microsoft have windows. Why? Because even in Seattle where it is often overcast and you seldom see the sun, people want to see the outside. There is a definite physiological and psychological need for fresh air and sun.

Many planners and architects say that it is physically impossible to give most people a window. Apparently these people have never stayed at a hotel where architects generally manage to put a window in every room.

Once again, this follows the guideline of imagining the home office. Virtually no one with a home office has it in a room without a window. (In my case my wife claims that I try to take the room in our home with the best view for my office—a disparaging falsehood.)

This means that as many people as possible have the added lift of seeing the outside world while working. And this increases both their happiness at work and their productivity.

Equally important, people know the company cares about and recognizes this need and is trying to accommodate its employees. Even if you don't get a window, you know that Microsoft at least made the effort to provide as many windowed offices as possible.

So who gets these windowed offices? (This is a big deal each time people are moved around because of the continuing change in group sizes—something that happens about every six months at Microsoft.) It's the one time seniority counts. Within each group, whoever has worked for the company the longest gets first pick, then the next longest, and so on. Seniority should count for something, but this is the only place it does at Microsoft.

This also stresses the egalitarian approach of the company because office assignment is not based on job level, but rather on seniority. (That said, at the VP level and above, their "group" always seems to be assigned a block of one to three offices, all of which have windows, so everyone ends up with one. However, I have seen people recently hired in at the director level not get a windowed office.)

> *Some companies actually have a policy that if an employee has a window and the person's job rating is not sufficient to warrant having the window, it will be blocked off. Not only does this reduce the productivity of that one employee, but it also delivers a clear message to every employee: The company will not do anything to improve working conditions until you reach a certain level. Better working conditions are a perk, and the higher your ranking, the nicer it will make it for you.*

SMALL BUILDINGS

Until Microsoft started growing beyond all reasonable expectations, buildings were kept small (around 200 offices) and designed so that over half of the offices had windows. The small building size allowed a more intimate environment to be maintained. The entire Win95 team had a building to itself so it could operate physically as a single, separate unit.

People do not want to be a tiny cog in a giant organization. Keeping the buildings separate and smaller gives people the feeling that they are a noticeable part of the effort in their building, and therefore in the entire company.

As Microsoft grew, it had to create bigger buildings, and it lost something in the process. Going to the larger buildings (that house several major projects) had a negative impact on morale for all employees, and it was particularly hard on those transferred into the larger structures. There was a definite sense that Microsoft had made the move from a focused, entrepreneurial organization to a huge corporation.

NO DRESS CODE

> *Microsoft rented out a number of available floors in the premier office building in downtown Bellevue, Washington, for part of PSS (technical support). Lessees in this building were primarily major law firms, accountants, and other expensive consultants.*
>
> *Riding the elevator became an experience for clients of these firms. Half the people would be dressed in very nice business suits and the other half in shorts and t-shirts. It became a daily reminder that dressing nicer is not indicative of a more profitable company.*

There is no dress code at Microsoft. And no dress code means exactly that. It's not polo shirts and slacks instead of suit and tie. During the summer, it's shorts and t-shirts, and every other week or so the *MicroNews* (the company newsletter) runs a warning that the health department requires that shoes be worn in the cafeterias. Barefoot elsewhere is accepted, and, in fact, I never actually saw anyone told to leave a cafeteria because of being barefoot.

Why is this so important? Well, if you were working at home, what would you wear? Employees can dress the same way at work. (I don't recall ever seeing anyone in a bathrobe, but it would have been acceptable.) Once again, this allows people to work comfortably and efficiently. Obviously, if employees prefer to wear a suit and tie or casual business dress because they feel more productive in "professional" attire—and there are a few who do—it's completely their option.

Even more importantly though, the freedom to dress as you wish delivers a very clear message: What is valued is your work; not how you dress. This is a crucially important message. People will respond to the priorities a company sets forth, and if how you dress matters, then some of the time and attention that would otherwise have gone to work instead goes into making sure you look good. Again, the goal is to keep employees focused on work and not diverted by irrelevant requirements.

Nothing is more demotivating than seeing someone who is less qualified advance because the person is a sharp dresser with a good head of hair. At Microsoft being a snappy dresser with a good head of hair is clearly not a requirement for advancement.

FREE SOFT DRINKS

Free soft drinks, coffee, juice, milk, and bottled water are provided in all buildings. Nowadays this has basically become a required perk at all software companies, and there is a good reason for this.

Once again, let's visit the home office. You walk into the kitchen, open the fridge and pull out a drink. So providing free drinks at the office makes it like home—better actually, because the drinks are there even if you forgot to go grocery shopping.

This again keeps the employee focused on work. Going down the hall to grab a pop and then returning to the office requires no break in concentration. But rummaging for change, asking to borrow a couple of quarters from a coworker, swearing at the machine because it took your money and failed to produce the drink—all of that breaks concentration.

The free drinks are also a constant reminder that the company cares. Yes, it's taken for granted after awhile. But it's still there, subconsciously at least, that the company finds you valuable enough to give you free drinks.

And finally, since most of the drinks have caffeine, sugar, or both, they keep people charged and blasting through their work. At 2 in the morning when debugging a killer problem, a cola every half an hour is the only way to keep going.

> *If you are in the lobby of any Microsoft building between 4:30 and 6:00 p.m., you will occasionally see someone's spouse and kids arrive to pick the employee up. The kids make a beeline for the kitchen asking over their shoulder if they can get a drink. The kids think that this is the best benefit in the world!*

OPEN SUPPLY ROOM

No keys, no forms, no tracking: Microsoft has an open supply room. Anything commonly used is available for the taking. Copiers are openly available too. If you need copies, you just run them.

For anything not commonly used, employees send an e-mail to "Supplies" and the requested items are delivered the next day. You don't need approval; it's just that things that are rarely needed aren't worth the shelf space in the supply room.

Now contrast this to a company that has a locked supply cabinet. What is this company saying? It is clearly saying that it does not trust its employees.

Employees will work to the expectations of the company. If the company says it doesn't trust them, then many will see no need to be trustworthy.

With all this talk about trust, come August when school is about to start, certain supplies do fly out of the supply rooms as employees stock up on their kid's school supplies. It happens to such an extent that some supplies are left in opened cartons in the rooms because they will be emptied out so fast.

This can basically be chalked up to an additional benefit one receives when working for Microsoft. And if it saves the employee a couple of hours of school shopping, then Microsoft is ahead of the game anyway in the extra work hours it got.

If supplies are open, then it is the employees' responsibility to not abuse that trust. If they are locked and you get something extra, well then you can take it home because it's the company's fault, not yours, that you got it.

This practice also, once again, keeps employees focused on work. Walking down the hall to grab some supplies is a minimal break from work. Filling out a form and waiting for supplies is a much bigger break and an annoying one at that. In addition, when the request is refused, and many must be if the system is there to safeguard pilferage, then that is even more annoying and even more of a distraction.

WORK ANY HOURS

There are no set work hours wherever possible. Of course, some positions have very set hours, such as the technical support staff and building receptionists, but most employees come in sometime before 10 a.m. and leave sometime after 4 p.m.

Please note: This is not the traditional flex time where employees can pick their hours but then have to be there those specific times every day. People can start at different times and work different hours each day.

So does this lead to people working a much shorter workweek? On the contrary, most people I know there work at least 10 hours each day. If you give people a set period to work, they'll work that period. If you tell them "whenever," they tend to put in a lot more time.

This also allows people to work according to their biological clocks—working longer days when they are in a very productive mode and working less when everything they do seems to go wrong.

Once again, this practice coincides with our home office reference point. When people work at home, they get up, eat breakfast, do one or two miscellaneous things, and then sit down to work. For some people, you could set your watch by when they sit down to work because their routine is so regular. For others, the time can vary by hours each day. And both work the time that is most effective for them.

So how do managers know what hours an employee worked? They don't. There is no tracking system, and at review time a manager has no knowledge of the hours each employee actually worked.

Employees are reviewed based on the work they accomplished. Nothing more, nothing less. This is an essential component to Microsoft's success: Employees are reviewed on their work—*and nothing else.*

This message reverberates every day as employees choose when to arrive at work and when to leave. The decision is based on what they need to accomplish that day in order to complete their project or meet the next deadline.

When a company institutes time cards, fixed hours, time clocks, or other tracking methods, it is saying that the amount of time spent at the office is more important than actually getting the job done. These devices may ensure a minimal level of productivity, but they decrease the actual productivity a company would otherwise have.

I had an office with a window overlooking a side entrance to the building. At about 12:30 each day, a group of us from within the building would go to lunch. I always knew when it was lunchtime because I could see one of my coworkers (name withheld to protect the guilty) running in to work to be there in time to join us for lunch. This same coworker is an extremely valued employee—based on his productivity.

MULTIPLE UNIQUE CAFETERIAS

Microsoft provides multiple cafeterias on-site, several with unique themes. The cafeteria in one building is almost entirely vegetarian, another primarily serves gourmet pizza, and yet a third serves ethnic dishes, rotating on a weekly basis. (Of course, a couple offer the standard burger/salad/sandwich fare as well.)

Why is this important? Mainly because employees do not need to leave the worksite to go eat. Driving offsite, waiting to be seated and then served, and driving back—this all takes time. By having food on-site, all of this extraneous time is eliminated and is instead spent working. An added benefit occurs when, as is frequently the case, teams that are working on a project or having an extended meeting "adjourn" to one of the cafeterias while still carrying on discussions. There is no break in flow as occurs when groups recess to go to restaurants for lunch.

But equally important, by providing a selection of unique cafeterias, work (and lunch!) is more enjoyable. Essentially Microsoft has a wide selection of different cafes. The food is good and the ambience pleasing. They are not four-star affairs, but they are much better than the single, large, one-size-fits-all, sterile cafeteria.

And like free drinks, the cafeterias provide an environment that is better than home because the selection is more varied and the food is prepared and ready for you.

It is also worth noting that there are no restricted cafeterias or dining rooms at Microsoft. Not even Bill has an executive dining room—he eats from the same cafeteria (although he usually eats at his desk, not in the dining room). Nor are there special dining rooms for customers who come to visit (although many times they will have food brought to a meeting room). Executive dining rooms, like other executive-only perks, are tremendously damaging to company morale. They deliver a clear message that only a few are important.

WRAPPING IT UP

Two common threads bind each of the aforementioned items. First, each was implemented to improve employee morale and hence increase employee productivity. Second, each item also provides a direct payoff by keeping employees focused on their job. Even if those items did not increase morale and productivity, they would still provide a direct return to the company: Employees put in more hours actually working, and they spend far more of their time focused on work and less on distractions.

And it all comes from Microsoft's understanding that employees are not a bunch of identical machines. Forcing all employees to fit the model of the "perfect" employee is like talking to a mule. It doesn't do any good and it annoys the mule.

By accepting that each employee has a different, unique way of working efficiently and by trying to give each employee the environment that is best for that person, Microsoft gets the maximum out of each employee.

Compare this with your average large company which is busy trying to force each employee into the same mold. All this does is lower morale, decrease productivity, and increase the amount of time the employee has to spend at work on items not relevant to the job.

In many companies it is almost as though management hates the employees and is going to make their life as difficult as possible. And management is definitely not going to try to make their time at work enjoyable—after all it's work damn it and it's not meant to be enjoyable.

Most companies do understand that a pleasant work environment is a giant bonus. That's why employees higher up the food chain are given "perks" like offices, windows, and special dining rooms. And most companies understand that these perks increase productivity. That's why they give them to senior executives. So the question is: Do companies think that most of their employees have different motivations than senior executives, or do they merely think that lower-level employees' productivity is not worth the expense?

Either way, companies are making serious, expensive mistakes by attempting to force individuals into predetermined molds and monitoring employees' every move. Think home office. Not implementing these items is managerial malpractice.

A WOLF
IN THE HENHOUSE

So what? Does it really matter if all companies are managed in a manner similar to Microsoft? What if they aren't? As the tagline for the movie *The Fly* (1986 version) said, "Be Afraid, Be Very Afraid."

Many small companies have had comparable management. What Microsoft did was successfully retain that management style within a large company. This has been largely due to the fact that no other company can buy Microsoft, since Bill and Paul Allen effectively own over 50 percent of the stock.

In the past, effective management systems have always been quashed when a large corporation acquired a small company and imposed its bureaucracy on the small company, eventually driving out all the exceptional employees. There now exists a very large company, run by a highly effective management style, and it cannot be bought out and ruined.

Microsoft's management has successfully combined most of the best aspects of both small and large companies. By operating as a big company where that is most effective and as a small company where that is appropriate, Microsoft has become a formidable competitor. And with its cash reserves and absence of debt, it can go after virtually any industry.

TROUBLE IN PARADISE

It is worthwhile to note that Microsoft is not perfect. In fact far from it. Microsoft engages in various practices that are, if not illegal, definitely not candidates for ethical-behavior-of-the-year awards. And Microsoft does intentionally do a number of things that many other companies choose not to do because they feel that those practices are wrong.

But to a large degree this is irrelevant. Microsoft's underhanded maneuvers give it tactical advantages at times but also make many companies unwilling to cooperate with it. In sum total, this does not result in an overwhelming advantage. If these practices did result in a significant advantage, we would have numerous successful companies who also utilize these same tactics.

OS/2 didn't lose out to Windows because Microsoft is unbeatable. It lost because IBM couldn't effectively develop or market PC software if its pants were on fire. Even Microsoft's core competency can be beat if done correctly.

The bottom line is that Microsoft's management techniques are what gives it it's competitive edge. And this is where opportunity exists for other companies. Because Microsoft could be just as successful, if not more so, if it stopped acting like a spoiled, arrogant teenager. More to the point, other companies can emulate the useful aspects of Microsoft's culture while retaining their integrity.

WE HOLD THESE TRUTHS . . .

The Declaration of Independence is a beautiful document that expresses not only the grounds upon which our forefathers elected to secede from England, but also the fundamental philosophy on which our government would be founded.

As a country we rarely live up to everything expressed in the Declaration, and in this we are no different from any other country. Where we differ is that we continuously strive to live up to it.

Microsoft is the same. In many areas, day to day, there are problems. Bureaucracy tries to insert itself into the system. Employees get too wrapped up in their own piece and don't take other projects into account.

The difference is that Microsoft unceasingly attempts to live up to its basic management philosophy. And it usually does so successfully. So, while Microsoft is imperfect, its goals remain paramount and mediocrity is not tolerated.

THERE ARE OTHERS

Other companies are following Microsoft's example. The companies in industries Microsoft is pursuing have to do the same to survive. Companies like Intuit and AutoDesk have had to develop management styles similar to Microsoft's to compete against it. And these companies too are setting their sights on other markets.

With the funding available to Internet and communication companies, there are now numerous other multibillion dollar companies that are also extremely effectively managed. These are small, highly productive companies that can react to changes within days. And these companies are going after existing, established markets.

This is similar to what happened 15 years ago when the economy became truly global. The effects of a global economy are still shaking out. Many companies found that they had competition at a level never before experienced, and they had to improve their own management or lose market share and possibly go out of business.

But the global change was easier. Countries erected trade barriers to give companies time to deal with the increased competition. Distance alone provided advantages to the local companies. And the difference was not that great between the effectiveness of management styles.

Microsoft and the companies like it have raised the bar again. Management in other companies must improve if they are to survive. But this time the bar is raised a lot higher than it was for global trade.

And this time the competitors are local and understand your market better than you do. There is no way to have outside influences slow them down. As the federal government is presently learning, even filing a lawsuit does not slow down Microsoft, or even divert it.

ANIMAL HOUSE

The critical question for many companies is: Can they change? At most large corporations, upper management operates like a frat house—a relatively friendly and clubby environment. Frats would never replace the president merely because someone else could do the job better.

In fact, a Bill Gates would never get promoted to a senior position in the average large company. Someone so nakedly aggressive would be recognized as way too dangerous and the entire herd would ensure his demise.

The most obvious proof of this? There has never even been a rumor of a large company like IBM or AT&T considering a senior Microsoft executive as its CEO. These companies know they need more effective management, but they are not willing to hire someone that focused on success.

Why not? Likely it's because the senior managers at these large companies realize that most of them would not survive if a Mike Maples or Steve Ballmer were made CEO of their company. The actions of these other large companies say more about their prospects against Microsoft than anything else. They realize that they cannot compete.

THE END

So what will happen? A few existing companies will effectively embrace enough of Microsoft's approach to management that they will survive. And some will even thrive.

A frightening number will become smaller and smaller each year, eventually going out of business or being acquired by one of the new companies coming along or by Microsoft itself.

That's the beauty of the free enterprise system. They'll go kicking and screaming. They'll ask for help from the government. But many large companies are facing extinction. And they can no more avoid their fate than the dinosaurs could.

So for many companies Microsoft simply means . . . the end.

INDEX

179

ABOUT THE AUTHOR

David Thielen has spent over 20 years on the bleeding edge of technology, including working at Microsoft as a senior software developer on Windows 95 and several other projects. Currently Director of Software Development for a hi-tech company, Thielen is the author of *No Bugs!* and *Writing Windows Virtual Device Drivers*. He has 22written articles for magazines including *The New Republic*, *PC Magazine*, and *Microsoft Systems Journal*.